The Holes in the Ozone Scare

The Scientific Evidence That the Sky Isn't Falling

The Holes in the Ozone Scare

The Scientific Evidence That the Sky Isn't Falling

by Rogelio A. Maduro and
Ralf Schauerhammer

21st Century Science Associates
Washington, D.C.
1992

In memory of
Warren T. Brookes

ISBN: 0-9628134-0-0

Library of Congress Catalog Card Number 92-64062

Front cover photo: L.A. Frank and J.D. Craven, University of
Iowa. Auroral oval surrounding North Pole, Feb. 16, 1982, imaged
in ultraviolet light by the Dynamics Explorer-1 spacecraft, traces
energy deposited into the upper atmosphere by solar wind
particles. A similar phenomenon exists at the South Pole, and the
so-called ozone hole occurs entirely inside that auroral oval.

Project editors: Marjorie Mazel Hecht and Christina Huth
Book design: World Composition Services, Inc., Sterling, Virginia

Please direct all inquiries to the publisher:
21st Century Science Associates
P.O. Box 16285, Washington, D.C., 20041
(703) 777-7473

Contents

Haroun Tazieff, former French secretary of state for the prevention of natural and technological disasters, is well known in France for exposing the lies of the catastrophists. Born in 1914 in Poland, his career has spanned many scientific fields from volcanic research, mineralogy, and geology to entomology. He is currently an elected member of his municipal and regional councils in France. He is shown here in June 1988 at Karakoram in the Himalayas.

This foreword was translated from the French by Rick Sanders.

Foreword

My first job was that of agricultural engineer. I worked at that from 1938 to 1939. The second, from 1945 to 1948, was as a mining engineer and geologist. The third is as a volcanologist, from 1948 to the present. And the fourth is being responsible for the mitigation of catastrophes, a political responsibility for the government of France from 1981 to 1986, and afterwards for the district of Isere in France.

I have burdened you with this curriculum vitae in order to specify upon what basis I am able to rationally approach new disciplines, to acquire in a few months or in a few years sufficient knowledge for reflecting upon and judging certain conclusions that some specialists, at times in peremptory fashion, put forward as demonstrated truths—when they are not. This basis is eight years of study at two successive universities, half a century of experience as an engineer and scientific researcher, and ten years of political responsibility.

If I permit myself this preliminary justification, it is because of a species of hysteria by which certain ecologists have been possessed in the last 20 or so years. This is a hysteria characterized by fear, sometimes panic, in the face of a civilization technically more and more sophisticated and to which, whether happily or deplorably, we belong. This fear leads the people in question to refuse everything implied in technological progress. Their war cry boils down to "No to ——." Replace the blank with high speed trains, supersonic airplanes, nuclear power, DDT, PCB, CFC, and even carbon dioxide (CO_2), and you will have the essence of what they are fighting against.

This self-justification is rendered necessary by the friendly accusations that these types, a certain species of ecologists, bring against me, namely, that of incompetence. According to them, strictly speaking, I could express my opinions on volcanic eruptions, forest fires, or earthquakes—but not on nuclear-generated electricity, nor the destruction of the ozone layer, nor dioxins, nor the greenhouse effect. Certainly, had I professed an alarm similar to theirs on these subjects, it would be much less likely that they would have accused me of "incompetence." On the contrary, they would have applauded me noisily. Unfortunately for me, I have had the audacity of denouncing and pointing out the deficits in the morality of their science, which, at times, are found behind vast, expensive, and consciously orchestrated catastrophist campaigns. Where does the money come from for this?

Here are two examples of articles that prompted my investigation of the subject: "Seveso: The Hiroshima of Chemistry," and "Seveso: 9 Months After: The Lessons Were In Vain." Both appeared in the popular science periodical *Que Choisir* [What Choice] in April 1977.

I was at the time a Citizen Lambda [John Doe], an individual among hundreds of millions targeted by the disinformation campaign launched on a global scale. I had believed in what was thus universally and imperatively affirmed as incontestable truth: that PCBs, and the dioxins they emit when heated to 300° Celsius, were frightful poisons. One or two years of this propaganda had led government officials—just as incompetent as I was in matters of polychlorobiphenyls—to make them officially illegal.

A half-dozen years later, I found myself responsible for the prevention of disasters, natural and technological, for the French government. The natural ones I knew quite well, since they are related to my profession. As for technological disasters, it was necessary to inform myself. The very first dossier I asked to have delivered to me—so much had I been convinced of the extreme hazard of PCBs—was the one on the explosion at the chemical plant in Seveso, Italy, in July 1976. The study of this dossier and the inquest I led at the time revealed to me, first of all, that this so-called catastrophe had had not one single victim. (This gives the "Hiroshima of Chemistry," as it had been baptized by an ostensibly serious monthly science magazine, a tinge of anticlimax.)

Second, I learned that dioxins, according to the judgment of *all* the actual experts consulted (and of the very knowledgeable Academy of Science), are not at all "frightful" and have never, anywhere, killed anyone.

Thus, the matter of presenting the industrial accident at the ICMESA factory in Seveso as an apocalyptic catastrophe was a matter of deliberate disinformation—in less diplomatic language what one calls a *lie*. The wherefore of this very expensive crime (just think of what is implied by a propaganda campaign on a global scale) is found in the shadowy concerns of monopolies. The production and commercialization of PCBs was coming to an end. These are substances, used as dielectrics, whose extraordinary qualities of chemical stability, nontoxicity, nonflammability, and nonexplosivity explain why they were being used by the entire world, with annual benefits that figure in billions of dollars. Those who held the PCB monopoly, rather than accepting the end of their control, decided to support a ban of their high-quality product and to find another patented molecule as a substitute.

As for the billions of dollars involved in the destruction of PCBs—which by now has become obligatory—and the cost of their replacement, this bill will be paid by the consumers and taxpayers—for that matter without their knowledge. If by chance they are informed of it, they do not rebel, since it is a question of "defending the planet."

Repeating the PCB Scenario with CFCs

The same thing that happened to PCBs seems to be happening now to chlorofluorocarbons (CFCs). This synthetic molecule, universally used in the refrigeration industry, in the packing industry, and in aerosol sprays, is of a quality just as extraordinarily excellent as the PCB molecule: CFCs are chemically stable, nontoxic, nonflammable, and nonexplosive. Now, for reasons that I do not know—I have not had the time to make the relevant inquiries—CFCs are the subject of a worldwide campaign blaming them for destroying the ozone layer.

The ozone layer contains barely a half dozen molecules of this substance, O_3, per million molecules of air (air is made up of about

78 percent nitrogen, N_2, 21 percent oxygen, O_2, and 1 percent of argon and other gases, of which 0.03 percent is CO_2). Nevertheless, they tell us, *without the least proof,* that this so-called layer of ozone protects us from the cancers that would be inflicted upon us by solar radiation. They tell us that CFCs, molecules much heavier than air, lift themselves up to the stratosphere, 10 to 15 kilometers high. There, because of the energetic solar ultraviolet, they lose the chemical stability that characterizes them within the biosphere and liberate their atoms of chlorine, which, they say, destroys the ozone.

On the other hand, the millions of tons of chlorine that are belched out annually by the volcanoes of the world (I know of what I speak: for more than 40 years my collaborators and I have been studying volcanic gas emissions) would have but a very secondary effect upon this destruction of the ozone, we are told, compared with CFCs, whose mass of chlorine is but infinitesimal compared with that of volcanic eruptions. The alleged harmlessness of volcanic chlorine is not invoked, it seems to me, simply to better incriminate the chlorine of the CFCs.

Whatever protection, thanks to stratospheric ozone, we may be presumed to enjoy, one question poses itself to my mind—a question for which the high-level chemists who accuse CFCs of destroying the ozone have not been able to offer a serious reply: How is it possible that nine-tenths of the CFCs called freons are made and used in the Northern Hemisphere—the hemisphere that is also the more populated and the more industrialized—yet it is in the Southern Hemisphere, which is for the most part uninhabited, that one sees the now famous "ozone hole"? And that this "hole" is not only in the Southern Hemisphere, but is above the largest and most total desert on the Earth, the Antarctic continent? The embarrassed replies I received were not satisfying at all.

It is because of this Antarctic "ozone hole" that I was Citizen Lambda with regard to these questions at the beginning of the campaign launched a few years ago, and as such put my whole faith in the enormous propaganda against CFCs. I began to reflect on this media drum-beat: TV, radio, dailies, weeklies, monthlies, popular science periodicals, the middle-level, high-level, and even the top scientific journals in the English language affirm these demonstrated truths.

Not only had the Antarctic interested me from childhood, but also I am quite familiar with it, having led four volcanological expeditions there between 1973 and 1979, right into the heart of the continent—near latitude 78, to the volcano Erebus, halfway between the magnetic and geographic poles, not simply skirting its edges as did some self-appointed Antarctica "experts."

Since this "ozone hole" was Antarctic, I was stimulated to get to the bottom of the question—of the why. The conclusion of my inquiry was that this hole was most probably in existence at the time of the Shackleton expedition in 1909, and that most probably, it is a natural phenomenon. Consequently, it has nothing to do with CFCs, which did not exist at that time. It would be too long to recount here the reasons that led me to the hypothesis that the "hole" has been in existence for many millennia, but since it is not yet more than a hypothesis (no more demonstrated than that this "hole" is the result of the action of CFCs), there is no point wasting time on that.

Scientific Bad Faith and the Ozone Hole

On the other hand, my inquest has led me to the conclusion that the alarmist hypothesis, according to which the CFCs are responsible for the Antarctic "hole," is based upon at least two acts of scientific bad faith, and that is sufficient to discredit it completely.

The first is the deliberate omission of all reference to previous publications bearing on the same subject: The propagandists have made us believe thus that the seasonal disappearance of the Antarctic ozone was discovered by a British researcher, J.C. Farman, *in 1985*. In truth, it was a score or so years earlier, in the 1950s— before CFCs were in widespread use—that the first and principal investigator of stratospheric ozone, Gordon Dobson, discovered that the Antarctic ozone faded away from August to November. (It is Dobson's name that is given to the units that measure the amounts of ozone in the ozone layer.)

The second infringement of scientific ethics was that in their powerful and sustained effort (crowned by success) to make an impression upon the good public—and this includes journalists, as well as people who occupy themselves with science and, above all, the political decision makers—the propagandists showed us satellite photographs of the Antarctic and of its "ozone hole" in the month of October only, which corresponds in the Southern Hemisphere to the month of April (that is, the beginning of spring). They did this without breathing a word to us about the fact that this "hole" shrinks from week to week as the Sun, which did not even make an appearance during the three months of winter, is rising on the horizon; without telling us that the higher the Sun is, the more powerful is the sum of solar energy that this part of the globe receives, and what's more, that this energy creates stratospheric ozone: From December on, this "hole" that the alarmists are using to sow panic, no longer exists.

This "hole" thus is a natural phenomenon, as Dobson had discovered before the existence of CFCs. Very likely it is due to the meteorological isolation that the Antarctic continent suffers each winter, an isolation that results in the enormous temperature gradient that marks the brutal limit between the extreme cold of the glacial ice cap and the more modest cold of the ice pack. On the one hand, the formidable glacial ice cap, 2,000 to 4,000 meters thick, covering the continent (and more), has temperatures on the order of $-80°C$. On the other hand, the ice pack is no more than a thin shell, hardly a few meters thick covering the *waters* of the ocean. Being liquid, these seas have a temperature near zero Celsius. This is an enormous gradient, 50 to 60 degrees Celsius, over a distance of hardly a few tens of meters.

On the contrary, in the Arctic the gradient is much, much weaker—of the order of barely $20°C$ over more than 4,000 kilometers. This is so for the very simple reason that the Arctic is an ocean, frozen on its surface only, which extends toward the south across the plains of North America and northern Siberia, where the winter temperatures are properly polar. This renders an extremely weak gradient: $10°$ to $20°$ over a number of thousands of kilometers, the opposite of the Antarctic, $60°C$ over a few hundreds of meters only.

Now, the temperature gradient between two masses of air conditions the direction and force of winds stirred up by differences of density, which were themselves generated by temperature differences. At high gradients—and there is none more marked than the one between the Antarctic continent and its ocean—there are the most powerful winds in the world.

That is why Antarctica is characterized by an extremely violent vortex of air currents, while a much weaker vortex exists in the Arctic. This winter vortex prevents the arrival above the Antarctic continent of air masses located at latitudes lower and ozone richer. They are richer because the Sun's radiation is not only more continuous there but also more effective (because of the inclination to the ecliptic of the Earth's axis of rotation) above Antarctica. The very cold and hence more dense air of the high altitudes thus falls downward, due to gravity, and empties the stratosphere of a good part of its summer ozone. Hence a possible explanation of why the tropospheric level of ozone in the Southern Hemisphere has been observed to be less than that of the Northern Hemisphere.

This so-called hole, generated during the winter, cannot be replenished with ozone during the long polar night, simply because of the lack of solar energy and the lack of tropical air. When day has returned but the "hole" is still at its maximum (in October, before the vortex completely breaks up, allowing ozone-rich air from the tropics to refill it, and before ozone begins to be generated by solar radiation reaching Antarctica), satellite detections of the "hole" are broadcast far and wide by the media and by concerned catastrophists.

Within little more than a month, there are no more photographs, because the famous "hole" shrinks, not only because of the arrival of ozone-rich air from the extrapolar regions, but also because of new ozone created from oxygen by the hard ultraviolet rays of the Sun. The mere fact that the propagandists do not show this rapid replenishment of the "hole," and that they do not say a word about it, reveals the doubtful faith hidden behind the colossal worldwide alarmist operation mounted to outlaw CFCs. The billions that such a campaign costs will be in any case recuperated from the consumer-taxpayers of the entire world, abused by this unedifying campaign of disinformation.

One could reply that I no more demonstrate the truth of my thesis on ozone than the alarmists do theirs. Nonetheless, it remains true that the ozone of the upper stratosphere is renewed unceasingly by the Sun at the expense of atmospheric oxygen. And the Sun will shine for another few billions of years. Hence to pretend that the stratospheric ozone is in the process of disappearing is, very deliberately, to say the opposite of the truth.

Here is the just-published statement of the general manager of Atochem, the largest CFC producer and the largest world producer of CFC substitutes (74,000 tons a year):* "The Rowland and Molina theory [of ozone depletion] is unscientific because it is based upon a model of a chemical reaction sequence without having proved the existence of the intermediary products; these reactions, which

* *Industrie et Environment,* No. 28 (June 26, 1991).

no one has ever reproduced in the laboratory, have never been observed anywhere."

The Unreal Greenhouse Scenario

As for the greenhouse effect, supposedly generated by CO_2 released by the burning of petroleum and its derivatives and from burning coal or wood, this seems to me also imaginary and just as unreal as the destruction of the ozone of the upper atmosphere.

The greenhouse effect induced by the atmosphere clearly exists. For that matter, without it there would be no more life on Earth than on the Moon. Without it, the temperature would be −150°C during the night and about 100° during the day. But to claim that some 3 and a few 10ths parts per 10,000 of natural CO_2 in the atmosphere will make the Earth's temperature increase reveals either naiveté or deceit.

CO_2 actually plays only a minor role in the greenhouse effect, the essential role being played by water, both in its visible form, little drops or crystals suspended in the clouds, and its invisible form, vapor. I consider as a proof of this that the greenhouse effect is maximal in humid regions and minimal in dry regions, while the proportion of CO_2 is exactly the same: 0.03 percent. Take a cloudless day (24 hours) in an equatorial zone and in a desert zone. The daily maximal temperature (in the shade) is of the order of 35 to 36°C in the Congo (for example) and 50 to 55° in the Sahara. The minimal nightly temperature is of the order of 28 to 30°C in the humid equatorial region, and from 0 to minus 5° in Tibesti or Hoggar; this is a half-dozen degrees of difference where there is 95 to 100 percent humidity, and about 50° where the humidity is not more than 20 to 30 percent. Where is the stronger greenhouse effect? Where water content is high.

Now the proportion of CO_2 is rigorously the same on the equator, in the tropics, at the poles, on the oceans, and at the top of mountains: It is not the carbon dioxide that determines the greenhouse effect; it is essentially the humidity of the atmosphere.

To claim that the increase of CO_2 in the atmosphere will drive up its temperature shows either an insufficient analysis of the causes of the greenhouse effect, or a certain bad faith. It is insufficient analysis, because one forgets that an eventual increase of the air temperature will increase both the evaporation of the waters of the surface of the Earth—primarily oceans—and the transpiration of plants. This will increase nebulosity, which will decrease temperatures during the day and increase them at night. Nebulosity, for that matter, will increase the albedo.of the planet; that is, the reflection into space of the solar energy received from the Sun. All of this implies a significant self-regulation of temperatures.

As for causing the polar caps to melt, let us have a good laugh! Their temperatures are some several dozens of degrees below zero, and it is not until you get above zero that ice is transformed into water.

The essential argument made by scientists in good faith—scientists who fear a fatal greenhouse warming from the measured increase of CO_2 over this last century—is the discovery made by Claude Lorius in the ice core samples that he drilled in Antarctic and Greenland ice caps. Lorius analyzed the microscopic fossil-air bubbles in the ice going back 150,000 years. His analysis shows that the curves of CO_2 content in the fossil air over 150 thousand years are parallel to the curves of the temperatures that prevailed at the moment the snowflakes were falling that were to be transformed little by little into ice. This shows that there was relatively little CO_2 during the glacials of the Quaternary Ice Age [the 2 million years of human existence] but very much during the interglacial periods, notably the last one, in which we are living, which is scarcely 8,000 years old. From this the conclusion is drawn that the more CO_2 there is, the warmer it is. Hence the catastrophic extrapolations based on the measured growth of atmospheric CO_2.

To this I counter that the opposite is just as credible with respect to cause and effect: To the now official thesis that the more CO_2 there is in the air, the warmer it is, I put forward the theory that the warmer it is, the more CO_2 there is in the air. To make my point, I submit that during the glacial periods, the extension of the glaciers reduced land area by about 40° of latitude (about 20° in each hemisphere); that is a surface at least on the order of 30 million square kilometers in each hemisphere. Since most of the CO_2 produced on Earth comes from either volcanic action or the biological kingdom, and the latter is absent on glacial caps, it is difficult to deny that the growth of atmospheric CO_2 corresponds to the retreat of these ice caps and the immediate recolonization by the biological kingdom of sea and land. Sixty million square kilometers of life makes a lot of CO_2.

The alternating of the glacial and interglacial periods, as Milutin Milankovitch demonstrated mathematically about 1930, is conditioned by three astronomical factors that vary in a regular fashion: By the inclination of the Earth's axis of rotation to the plane of the ecliptic, whose period is on the order of 40,000 years; by the eccentricity of the Earth's orbit in its relation to the Sun, of a period of about 100,000 years; and by the precession of the equinoxes, 20,000 years. The quantity of solar energy that the surface of the Earth receives, and hence its climate, depends almost exclusively on these factors. The alternating advance and retreat of the glaciers and the duration of these epochs depend on the interaction of these three factors.

Apart from these major variations, climate is sometimes heavily influenced by the activity of the Sun itself, which is in part cyclic (the 11-year sunspot cycle is the best known) and in part apparently unpredictable—but definitely important.

On the other hand, the importance of man-made CO_2 in the evolution of climates and the advance or retreat of glaciers is, at present, just as insignificant as the effect of man's coastal activity upon the daily tides, or that of industry upon the force of the winds, the unleashing of a hurricane, or the trajectory of a cyclone.

Windmills for Ecologists

The normal citizen, which I believe myself to be—one who in good faith believes what the media proclaim, especially since the media are all in agreement—finds himself obliged to reflect upon the claimed disappearance of stratospheric ozone, the claimed increase of the greenhouse effect by man-made CO_2, and the alarmist "arguments" that have implicated the Antarctic: the "ozone hole" and the melting of the ice cap. Now, the Antarctic is close to my heart, especially since I have led several volcanological expeditions to the heart of this continent. And this reflection brought me finally to formally question the alarmism that is today the fad—alarmism that brings into play billions of business dollars per year (and that might explain certain positions taken), alarmism that sweeps up those who merchandise news, since the bad sells incomparably better than the good.

The unleashing of the media, first against PCBs, then CFCs, and then CO_2, puts into panic the ecologists, who are always ready to see the devil behind modern technology. It also puts into panic voters of every stripe, dragging in the political decision-makers (at least those who had not already made up their minds after being sensitized by lobbies whose interests in the matter are far from negligible). And thus it is that substances incomparably less dangerous than thousands of others that are daily and widely used have been banned from society: first PCBs, then CFCs, and finally CO_2. The whole world loses in this—except for those "happy few" who gain from it. Thanks to them.

Here I have summarized the quite solitary path that I followed in discovering that the catastrophes announced by great blowing of the trumpets are no more than windmills for naive ecologists to tilt at. After I took my stand, first in a little book published in 1989, then in public debates and radio and television appearances, I had the surprise—oh, how pleasant—of receiving the approbation of numerous scientists, especially specialists in these matters. On the other hand, I have attracted innumerable enmities, some naive and some from certain people of doubtful honesty. But a small number of friends of quality is worth more than a bunch of fans or a bunch of foes.

—*Haroun Tazieff*

The Holes in the
Ozone Scare

THE EARTH'S ATMOSPHERE

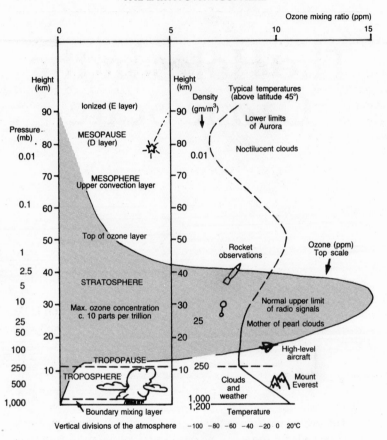

Source: Adapted from Hugh Ellsaesser.

Introduction

"**T**he sky is falling! The sky is falling!," cries Chicken Little when an acorn hits him on the head. Chicken Little spreads the alarm to Henny Penny, Turkey Lurkey, and others, who accompany him to tell the King. On the way, they meet up with Foxy Loxy. In the guise of showing them a short cut to take them straight to the King, Foxy leads them right into his cave—and into his dinner pot. So ends the original version of the traditional children's story.

Had even one inhabitant of this storybook kingdom asked Chicken Little for some scientific evidence, the fox would not have had such a magnificent meal.

This book is intended as a guide for citizens who want to ask that important question: Where is the scientific evidence? Is the sky really falling? Is the ozone layer really being depleted by chlorofluorocarbons (CFCs)—or are we victims of a sophisticated Foxy Loxy?

This is exactly the situation the authors faced in 1988. One of the authors (Roger Maduro), then a believer in the ozone depletion theory, was gathering evidence to write an article to show that the global warming theory was a scientific fraud. In the course of an interview with Reid Bryson, head of the Institute for Environmental Studies in Madison, Wisconsin, the author was startled to be told

1

not to pay attention to the ozone depletion theory because there was a volcano in Antarctica that pumped more chlorine into the atmosphere than did the entire annual production of CFCs on Earth. If chlorine were indeed responsible for the Antarctic ozone hole, then it would be this volcano that was responsible, not CFCs, said Bryson.

This information set the author, a geologist by training, on a quest. A few phone calls later, the top volcanologists in the United States had confirmed what Bryson had said: Mt. Erebus in Antarctica pumps more than 1,000 tons of chlorine a day into the atmosphere. This meant that Mt. Erebus was lofting more chlorine into the atmosphere in one week, than an entire year's production of CFCs!

This book is the result of that quest for the scientific truth about ozone. Because the material presented will challenge most readers' basic assumptions, we have quoted extensively from the many scientists and scientific papers that demonstrate why the ozone depletion theory is a scientific fraud. Most of the evidence presented here can be confirmed by a trip to a library that subscribes to scientific journals.

As readers will see, the new twist to the Chicken Little story is that today Chicken Little and Foxy Loxy seem to be working together to convince people that the sky is falling. The giant chemical corporations, which are slated to make hundreds of billions of dollars selling replacements for the now-banned CFCs, are working with the environmental movement, which has already made millions of dollars in revenues from the ozone depletion scare. The U.S. environmental groups are able to finance the promotion of environmental hoaxes like ozone depletion through the more than $500 million they receive a year from the major philanthropic foundations run by this nation's financial elite—Rockefeller, Ford, MacArthur, and other foundations. But money is not the only motive driving the ozone hoax. Behind the actions to ban CFCs— and to cut back on refrigeration—is the Malthusian ideology that the world needs fewer people.

The cost to the public of banning CFCs and other beneficial chemicals will be even greater than the enormity of the profits for those who will replace the chemicals. The next time the headlines blare, "the sky is falling," the public should ask, "cui bono?"—who benefits?—and start looking for those acorns.

What Are CFCs?

To understand the importance of CFCs, we have to look at their history. The advent of electricity at the end of the 19th century dramatically changed how people lived. Parallel to the rapid spread of electricity and the scientific, technological, and medical discoveries that occurred around the turn of the century was a rapid rise in the lifespan of the American and European populations. Electricity provided power not only for lights, but also for entirely new technologies like food refrigeration, for individual households and for industry and agriculture. Foods that would previously spoil in the heat were now available in abundance at all times of the year and could be shipped under refrigeration. Electricity made it possible for ordinary citizens to have refrigerators in their homes.

There was a major problem, however, in terms of the safety of the refrigerants used. In 1928, the most common refrigerants were sulfur dioxide (which held most of the market), methyl chloride, and anhydrous ammonia. From a mechanical standpoint, sulfur dioxide was a good refrigerant but it was very corrosive to the metal parts of the refrigerator; methyl chloride was more expensive, but not nearly as corrosive to metals. From the human standpoint, however, these chemicals were all highly toxic. Sulfur dioxide and anhydrous ammonia, the most toxic, have a pungent odor sufficiently noxious to awaken a sleeping person. The other coolants, however, did not have a strong odor to warn of their presence. It became common, therefore, for entire families to be killed by the release of toxic gases as a result of a refrigerator leak.

The danger of the existing refrigerants was dramatically demonstrated in 1929, when more than 100 people died in a Cleveland hospital from a leak in the hospital's refrigeration system. *The New York Times* and other newspapers were waging a major campaign to ban household refrigerators as too dangerous, and fears about poisoning were the major reason at that time that 85 percent of U.S. families with electricity had no refrigerators in their homes. As a result of these disasters and the publicity, the future of refrigeration was at stake.

At that time, Frigidaire, a subsidiary of General Motors, was the largest manufacturer of household and light commercial refrigerators. In 1928, a few months before the Cleveland disaster, Frigidaire had decided that household refrigeration in the United States would not have a promising future until a nontoxic refrigerant was invented. A team of chemists led by Thomas Midgley at General Motor's corporate research laboratory was assigned to find this refrigerant.

Midgley and his associates, Albert Henne and Robert McNary, all of them brilliant chemists, began an intensive search of potential candidates in the periodic table of the elements. They eliminated chemicals that were unstable, insufficiently volatile, or with too low a boiling point. Their search left them with eight elements that could work: carbon, nitrogen, oxygen, sulfur, hydrogen, and the halogens (fluorine, chlorine, and bromine). Since refrigerants based on the first five chemicals would likely be toxic or flammable, Midgley and his associates ruled them out. The halogens, however, had excellent nonflammable and nontoxic properties.

The chemists hypothesized that a mixture of a nontoxic but flammable chemical with a nonflammable but toxic one would perhaps become nonflammable and nontoxic. After an exhaustive study, Midgley's team settled on fluorocarbons and proceeded to synthesize these new molecules. Albert Henne had done his doctoral research on fluorocarbons, compounds that are similar to hydrocarbons with the exception that one or more atoms of hydrogen are replaced by a chlorine, fluorine, or bromine atom. These compounds, found naturally in seawater, are also chemically inert and stable, another critical factor in the search for the ideal refrigerant. In less than two weeks they had synthesized the first chlorofluorocarbon (CFC), a fully chlorinated fluorocarbon. All hydrogen atoms in the hydrocarbon chain had been replaced by either fluorine or chlorine.

The discovery was not announced to the public at the time because of the hysteria against refrigerators. Frigidaire first wanted to ensure the safety of the new compounds, and so two years of testing and research followed, during which time Midgley and his associates synthesized several more CFC compounds.

The introduction of CFCs to the public in 1930 could not have been more dramatic. Addressing a packed meeting of the American Chemical Society, Thomas Midgley took a container full of CFCs, put the open top against his lips, and inhaled deeply—to the astonishment of the audience. Then he turned around and slowly exhaled in front of a candle, extinguishing it. A new refrigerant was born.

CFCs and the Food Supply

CFCs soon proved to be the ideal refrigerants. Nontoxic, non-flammable, cheap, simple to produce, and extremely stable and unreactive, the compound quickly produced a major change in the United States and the rest of the world. Household refrigeration became a possibility for the majority of the population, thus up-grading the standard of living.

Today, chemical companies are having nightmares attempting to replace CFCs. As Midgley discovered during his exhaustive work, there are just so many chemicals in the periodic table, and the fluorocarbons happen to be the only ones that meet the criteria. It is for this reason that despite multi-billion-dollar research efforts, the giant chemicals cartels still have not found good nontoxic replacements for CFCs, and the likelihood is that they will not. Any breakthrough in refrigeration will come only from the discovery of some new type of machinery or refrigeration system, something that may take decades to implement. What we see today, as a result of the CFC ban, is that in many cases the refrigerants replacing CFCs are the same toxic chemicals that CFCs were created to replace in 1929!

The issue of refrigeration is covered extensively in Chapters 8 and 9. Just as the ozone depletion propagandists and the news media never mention the scientific evidence refuting their theory, they never mention the consequences of the CFC ban. As world-wide refrigeration experts have been warning, the ban on CFCs will collapse the worldwide cold chain that provides for the storage and transportation of food. Their conservative estimates are that between 20 and 40 million people are going to die from starvation and food-borne diseases every year because of the collapse of refrigeration. It is estimated that an additional 5 million children will die from lack of immunization: Vaccines have to be refriger-ated, and the ban on CFCs will prevent Third World countries from building the units required to store the vaccines in remote areas.

The Ozone Depletion Story

The theory that man-made CFCs would deplete the ozone layer is only one of many theories claiming that ozone depletion would lead to doomsday. The theory originated in March 1971, when James McDonald, an atmospheric physicist from the University of

Arizona, testified at congressional hearings on the Super-Sonic Transport (SST) program. At the time there was a major fight to kill the SST program, but all of the arguments of the opponents had failed so far. McDonald's testimony centered around his theory that water vapor emissions from the SST were going to wipe out the ozone layer, allowing a large amount of ultraviolet radiation to penetrate the surface of the Earth, which would allegedly cause a massive increase in skin cancer incidence. As elaborated in Chapter 2, the news media seized upon the skin cancer story and made it the issue of the day. Funding for the SST was killed, and the ozone depletion theory was born.

(McDonald, it should be noted, had previously testified in Congress as an ardent proponent of the theory that UFOs—unidentified flying objects—regularly visited the Earth, causing major electrical blackouts in the process of recharging their alien spacecraft.)

Once the skin cancer scare had been established as an issue that would get the news media's attention, ozone depletion theories began to proliferate. These theories maintained that the ozone layer was going to be wiped out by nitrogen oxides (rather than water vapor) from SST exhausts, by nitrogen oxides from atmospheric nuclear tests, by nitrous oxide from nitrogen fertilizer, by chlorine from the Space Shuttle exhaust, and by emissions from pesticides, fumigants, and what not.

The theory claiming that CFCs would deplete the ozone layer was theory number 5, invented by F. Sherwood Rowland and Mario Molina in December 1973. F. Sherwood Rowland was then head of the chemistry department at the University of California at Irvine and Molina was his assistant. At the time, the first three ozone depletion theories (SSTs, atmospheric nuclear tests, and fertilizers) had faded into the background. The theory in vogue was that chlorine from the Space Shuttle exhaust would wipe out the ozone layer. Rowland and Molina, however, found a much better source of chlorine in the stratosphere than the Space Shuttle—CFCs.

Rowland and Molina's Theory

The Rowland and Molina theory says that CFCs are so inert that there are no sinks (nothing to capture or destroy them) in the troposphere (the portion of the atmosphere below the stratosphere). Therefore, CFCs have very long lifetimes in the atmosphere. According to the theory, the most common CFCs, CFC–11

THE VILLAIN: A CFC-12 MOLECULE

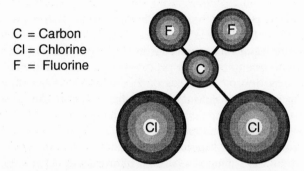

C = Carbon
Cl = Chlorine
F = Fluorine

The chlorine atoms in this schematic of CFC-12 are the alleged villains set loose to destroy ozone molecules.

and CFC–12, both very long lived, remain in the atmosphere about 75 and 120 years, respectively. After 5 years in the troposphere, the CFCs are transported into the stratosphere. There ultraviolet rays break them up into "free" chlorine atoms (those that can combine with other elements) and other molecules. This free chlorine then breaks down ozone molecules.

Specifically, according to the theory, the following reaction is alleged to happen to the CFC–12 used in household refrigerators. CFC–12, or CCl_2F_2) undergoes the following chemical reaction:

$$CCl_2F_2 + \text{ultraviolet radiation} \rightarrow Cl + CClF_2.$$

The single chlorine atom (Cl) then combines with an ozone molecule (O_3) to form a chlorine monoxide molecule (ClO) and molecular oxygen (O_2):

$$Cl + O_3 \rightarrow ClO + O_2.$$

The chlorine monoxide molecule left by this step is also quite reactive and, according to Rowland and Molina's claims, it quickly combines with atomic oxygen (O) in the stratosphere to release another oxygen molecule plus more atomic chlorine:

$$ClO + O \rightarrow Cl + O_2.$$

To sum up the Sherwood and Molina disaster theory, we quote from a July 1988 article in *Physics Today*: "The net result is that ozone molecules are removed from the stratosphere and chlorine atoms are free to begin the process over again. A single chlorine atom may destroy hundreds of thousands of ozone molecules during its residence in the stratosphere. This reaction cycle is interrupted when the free chlorine atoms become sequestered in so-called reservoir compounds."

From this purely hypothetical beginning spring major catastrophe theories of allegedly harmful ultraviolet radiation wreaking destruction on Earth.

The Evidence

Fortunately, Rowland and Molina's version of atmospheric chemistry is not the whole story, nor are the various ultraviolet radiation catastrophe theories.

CFCs are inert, nonreactive, nontoxic, nonflammable chemical compounds that do not destroy ozone or anything else. Omitted from the hypothetical stories of CFC's mass destruction of ozone is the fact that the amounts of chlorine contained in all the world's CFCs are *insignificant* compared to the amount of chlorine put in the atmosphere from natural sources (Chapter 1).

Further, there has yet to be published a single scientific paper that presents any documented *observations* of CFC molecules actually breaking up in the stratosphere. The chemical reactions described by Rowland and Molina have been carried out only in laboratory experiments. Rowland and Molina have based their *theoretical model* on just a few chemical reactions in a carefully controlled laboratory setting. In the real world, at least 192 chemical reactions and 48 photochemical processes have been observed to occur in the stratosphere. Most of these reactions are very fast processes involving highly reactive species, particularly free radicals and atoms in excited states, whose reactions can affect the chemistry of the stratosphere at very small concentrations. These reactions are extremely difficult even to reproduce in the laboratory; measuring their rates would be yet more difficult.

Some scientists have challenged Rowland and Molina's laboratory experiments. One of the criticisms is that they carried out their experiments of CFC photolysis in the laboratory with the gas confined in glass tubes and that they disregarded the possible edge effects in these tubes that can greatly distort results.

Lastly, CFC molecules have never been observed to rise as high as 40 to 60 kilometers (30 miles), the only altitude at which the ultraviolet radiation is intense enough to break up CFC molecules. Present claims are based solely on the supposition that the heavier-than-air CFCs *will* rise to the stratosphere *because* they are not water-soluble molecules, which means there are allegedly no sinks—or resting places—for them on the surface of the Earth.

To take a couple of reactions involving just a few molecules, carry them out in an isolated laboratory environment, and then claim this is what happens in the stratosphere (where it cannot be measured) is scientifically preposterous. For this reason, Rowland and Molina carefully prefaced their 1974 ozone scare paper with the following disclaimer: "We have *attempted* to calculate the *probable* sinks and lifetimes for these molecules" [emphasis added]. Such disclaimers, however, are never mentioned by the press; instead, a theoretical model is reported as observed fact.

This book aims to provide the scientific evidence that will enable the reader to make his own informed judgment on the issue. Although there is great deal of scientific detail in the book, we have written the book for the layman. The detail was necessary because the proponents of the ozone depletion theory have deliberately obfuscated the facts about ozone research and omitted the most critical factors that would enable an informed citizen to make his own judgment based on the evidence.

Beyond the work that we present in the book, we have gone a step further and enlisted the help of the great pioneer of ozone research, Gordon M.B. Dobson. There is no more crushing refutation of the ozone depletion theory than Dobson's writings. Born in England in 1889, Dobson became the foremost researcher of the ozone layer in this century and remained so until his death in 1976. As a result of his contributions, the units that measure the thickness of the ozone layer were named after him.

As a lecturer at Oxford University, Dobson's qualities were manifested not only in his great scientific discoveries but also in his special ability to inspire his students and his audience. His many students and collaborators have become some of the leading figures in atmospheric sciences today. Dobson's ability to excite his audience with the beauty of science and to make complex subjects understandable is reflected in his 1968 book, *Exploring the Atmosphere,* a classic work in atmospheric science. We have reprinted Chapter 6 from his book, "Ozone in the Atmosphere," as an appendix beginning on page 323.

Descriptions of the ozone layer found in the press today are so incompetent as to make it impossible for even the best-educated reader to determine what is really going on. In contrast, Dobson's chapter reveals the complexities and behavior of the ozone layer in language understandable to the layman. Dobson's description of the dynamics and chemistry of the ozone layer by itself should enable the reader to judge why the ozone depletion theory is a fraud.

In the first six chapters of this book, we present a point-by-point refutation of the today's ozone depletion theories. In the next five chapters, we answer the question *qui bono?*—who benefits from the banning of CFCs and other useful chemicals as a result of the ozone depletion fraud. Finally, in the last chapter, we present a comprehensive proposal to reverse today's actual environmental disasters that result from the collapse of the world economy: famine, starvation, epidemic disease, and industrial shutdown.

References

Mario J. Molina and F.S. Rowland, 1974. "Stratospheric sink for chlorofluoromethanes: chlorine atomc-atalysed [sic] destruction of ozone," *Nature,* Vol. 249 (June 28), pp. 810–812.

1

Natural Sources of Chlorine Are Much Greater Than CFCs

The ozone depletion theory does not claim that CFCs deplete ozone; it claims that a chlorine atom released by the breakdown of CFCs depletes the ozone layer. If it were true that chlorine from CFCs could wipe out the ozone layer, then Mother Nature would appear to be suicidal. Chlorine is one of the most naturally abundant trace chemicals in the atmosphere. The natural sources of chlorine in the atmosphere dwarf the tiny amounts of chlorine that could possibly be released by all the CFCs on Earth. Based on the evidence, in fact, if chlorine is truly a threat to the ozone layer, then the government should cap volcanoes and prohibit seawater from evaporating.

The yearly production of CFCs today is estimated at approximately 1.1 million tons, which includes approximately 750,000 tons of chlorine. Compare this to the natural sources of chlorine gases, as shown in Table 1.1 and Figure 1.1.

• More than 600 million tons of chlorine are released into the atmosphere every year by the evaporation of seawater, which contains salt (sodium chloride, NaCl). Although most of this chlorine is washed out by precipitation, large amounts of it still reach the stratosphere, through the pumping action of thunderstorms, hurricanes, typhoons, and other cyclonic activity.

11

Table 1.1
ATMOSPHERIC SOURCES OF CHLORINE
(millions of tons per year)

Seawater	600.0
Volcanoes	36.0
Biomass burning	8.4
Ocean biota	5.0
Total natural sources	**649.4**
Chlorine in CFCs	0.75
Chlorine theoretically released by the alleged breakup of CFCs	0.0075

• Passively degassing volcanoes pump more than 36 million tons of chlorine gases into the atmosphere in ordinary years when there are no volcanic eruptions. Great volcanic eruptions pump from a few million to hundreds of millions of tons of chlorine into the atmosphere. Most important, violent volcanic eruptions will inject gases and debris directly into the stratosphere.

• There are 8.4 million tons of chlorine gases produced by forest fires and the burning of biomass, largely as a result of primitive slash-and-burn agriculture methods and the lack of modern energy sources in the developing sector.

• Ocean biota, including algae, kelp, and plankton, have been measured to emit more than 5 million tons of methyl chloride into the atmosphere, and vast amounts of this biotic source of chlorine have been measured high in the stratosphere. Recent studies indicate that land plants may also contribute vast amounts of methyl chloride to the atmosphere.

• In addition, untold numbers of tons of chlorine enter the Earth from outer space, a result of meteorite showers and cosmic dust burning up as they enter the atmosphere.

These comparisons are even more startling when the amounts of chlorine allegedly released from CFCs are compared to natural sources of chlorine. According to the theory, approximately only 1 percent of the CFCs produced on Earth is broken up in the stratosphere every year. (The reason is that CFCs, because they are chemically inert, have lifetimes of more than 100 years in the atmosphere).

Therefore, a year's production of CFCs would contribute *at most* 7,500 tons of chlorine to the atmosphere. In plain English, the amount of chlorine contributed to the stratosphere by CFCs is barely one-tenth of 1 percent the amount of naturally occurring

Figure 1.1
ATMOSPHERIC SOURCES OF CHLORINE
(millions of tons)

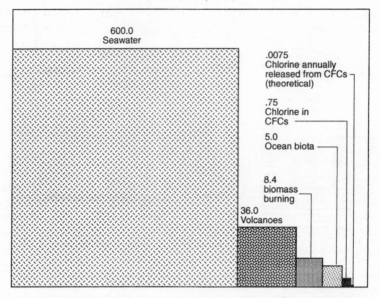

chlorine. This, of course, rests on the assumption that CFCs are being broken up in the stratosphere, an assumption for which there is no observational evidence.

Antarctic Chlorine and the Ozone Hole

One of the only documented facts among the ozone scare stories is that the concentration of active forms of chlorine in the region of the so-called Antarctic ozone hole is 100 to 1,000 times greater than that at the same level in adjacent areas. Therefore, the propagandists conclude, CFCs are arriving at the South Pole in great concentrations and are being broken down by ultraviolet radiation, releasing the killer chlorine molecules that then poke a hole in the ozone layer.

This Antarctic chlorine comes only from CFCs, the ozone hole propagandists say, using this as the final proof that the Rowland and Molina theory of ozone depletion is correct. Ignored by the ozone depletion theorists is the fact that less than 10 km (6 miles) upwind from the observation station at McMurdo Sound, Antarc-

13

Guilty of emitting 1,000 tons of chloride per day into the stratosphere: Mt. Erebus in Antarctica, located just 10 kilometers upwind of Mc-Murdo Sound, where the ozone measurements are taken.

tica, where the chlorine concentration readings are being taken, there is a volcano, Mt. Erebus, which began an active cycle of eruptions in 1972. Unlike most volcanoes, Mt. Erebus does not erupt and then go into a period of inactivity. Instead of having a lava caldera 2 or 3 km deep below the volcanic cone, the caldera in Mt. Erebus is on the surface, which means that the volcano is constantly erupting.

From observations made in 1983 by volcanologist William Rose of Michigan Technological University, published in *Nature* magazine, it has been estimated that Mt. Erebus ejects more than 1,000 tons of chlorine a day into the atmosphere. This comes to about 370,000 tons of chlorine a year, which alone represents almost half of the entire world's production of chlorine for CFCs (about 750,000 tons). Mt. Erebus's output is even more dramatic compared to the amount of chlorine released by the alleged breakup of CFCs in the stratosphere (7,500 tons). All by itself, Mt. Erebus, pumps 50 times more chlorine into the atmosphere annually than does an entire year's production of CFCs (Figure 1.2).

In short, the chlorine measured in Antarctica should be no mystery. Mt. Erebus is constantly blowing out a huge cloud of

Figure 1.2
MT. EREBUS CHLORINE OUTPUT COMPARED WITH CHLORINE RELEASED FROM BREAKUP OF CFCs
(tons)

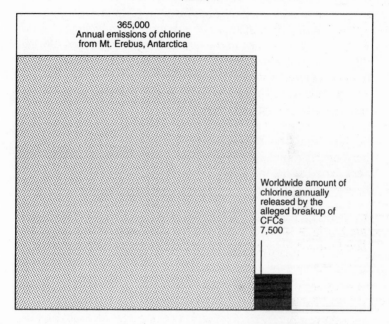

365,000
Annual emissions of chlorine
from Mt. Erebus, Antarctica

Worldwide amount of chlorine annually released by the alleged breakup of CFCs
7,500

chlorine and other volcanic gases; the wind picks it up and carries it a mere 10 km downwind. There the scientists at McMurdo Sound measure it with very expensive equipment on the ground and with very sophisticated equipment sent aloft with balloons. The balloons go straight through the volcanic cloud. Most astonishing, however, is that not a single report from these scientists has even mentioned the existence of this volcano. The public is led to believe that all this chlorine in Antarctica comes from CFCs.

Mt. Erebus is of even more interest in the ozone story. It is located along the area of the fierce stratospheric polar jet stream. With its high altitude—13,000 feet—a large portion of the hundreds of thousands of tons of chlorine spewed out by Mt. Erebus is picked up and advected downwind by the jet stream, which then carries the chlorine thousands of kilometers as it circles the pole. It is this swift jet stream, measured to flow as fast as 300 miles per hour, that creates the polar vortex. This vortex seals the atmosphere over the entire continent of Antarctica for almost two months of the year, creating the dynamic conditions for the natural cycle that leads to the thinning of the ozone layer over Antarctica for a period of 30 to 60 days of the year, the so-called ozone hole. (This will be addressed in more detail in Chapter 5.)

The chlorine from many active volcanoes has a lifetime of a few weeks to a few months in the atmosphere because the chlorine dissolves in water and returns to the ground with rain. The Antarctic atmosphere, however, is extremely dry, so that the chlorine compounds from Mt. Erebus have longer lifetimes in the atmosphere before being precipitated.

There is no evidence that any of this chlorine found at McMurdo Sound comes from the breakup of CFCs. Sunlight reaches Antarctica only six months of the year and for the other six months it is very dark. During the height of the Antarctic summer, the Sun is barely a few degrees above the horizon. In fact, the ultraviolet radiation that reaches the Antarctic stratosphere for most of the year is very weak because it has to travel great distances horizontally through the rest of the atmosphere. It does not have the intensity necessary to break up the strong bonds in CFC molecules at this level.

The Role of Volcanoes and Chlorine Worldwide

The case of Mt. Erebus brings us to the much broader issue of global fluxes of chlorine.

How much chlorine is injected into the atmosphere from volcanoes and how much of this makes it into the stratosphere? In the 1970s, the standard estimate for global emissions of chlorine from volcanoes was 7.6 million tons per year, according to calculations of volcanologist O.G. Bartels, published in 1972. Some volcanologists argued that volcanoes emitted much greater amounts of chlorine, but there was little hard evidence. The problem in measuring chlorine from volcanoes is twofold. First, chlorine in the atmosphere, like many other gases, is extremely difficult to measure. Second, getting close enough to a volcano to measure what is coming out of it is a very risky proposition. Therefore, all present-day estimates of volcanic fluxes are "educated guesses."

One of the procedures for measuring the amount of chlorine and other gases emitted by volcanoes is to estimate the chemical composition of the hot magma before an eruption and then to examine the chemical composition of volcanic rocks left after the lava has cooled. Scientists then calculate the difference in the chemical content and multiply it by the amount of lava calculated to have erupted from the volcano; the result is the estimate of gases released by an eruption.

Another procedure is to analyze samples taken from the smoke plume. This is so difficult to measure in the case of chlorine that usually scientists measure the sulfur and then calculate the amount of chlorine emissions as a ratio to the sulfur emissions.

The first systematic measurement of chlorine gases from volcanic eruptions was carried out at the end of the 1970s by a volcanologist with the U.S. Geological Survey, David A. Johnston. In a July 1980 paper published in *Science* ("Volcanic Contribution of Chlorine to the Stratosphere: More Significant to Ozone Than Previously Estimated?"), Johnston states that the actual amounts of chlorine released by volcanoes may be 20 to 40 times greater than previously estimated and vastly more than all the chlorine contained in CFCs. Even these estimates may be very low, Johnston says:

> ... [I]f the rate of degassing beneath the surface is rapid in comparison to the rates of magma ascent and eruption, then gas may have escaped in early explosions from magma that did not erupt until later or perhaps did not erupt at all. In that case, the total atmospheric and stratospheric HCl [hydrogen chloride] injections may have been higher than these estimates" [p. 492].

Specifically, he says, one single volcanic eruption in 1976 put more chlorine into the atmosphere than the entire amount of chlorine contained in the CFCs manufactured in 1975.

Until Johnston's published work, the chlorine output of volcanoes had been estimated by assuming that magma contains 0.02 to 0.025 percent (by weight) chlorine before eruptions, and that this was all released during the eruption. In his more careful measurements, Johnston found the percentage of chlorine in the magma to be actually 20 to 40 times greater—0.5 to 1.0 percent. Discussing the effect of volcanoes on the ozone layer and climate, Johnston says that major eruptions may have a long-term impact upon stratospheric ozone. He cites as an example the eruption of the Bishop Tuff from Long Valley Caldera, California, 700,000 years ago. This eruption may have injected 289 million metric tons of HCl into the stratosphere, he says, the equivalent of about 570 times the 1975 world industrial production of chlorine in fluorocarbons! "Clearly, volcanic sources of stratospheric chlorine may be significant in comparison with anthropogenic sources," Johnston says.

Johnston had intended to measure the actual emission of chlorine from volcanoes worldwide but, unfortunately, he died at his observation post when Mt. St. Helens erupted in 1980 in the state of Washington.

What are the global fluxes of chlorine from volcanoes? Using the more detailed chlorine analysis made by David Johnston and multiplying by the previous estimates (7.6 million tons) produces an estimate that the annual emission of chlorine from volcanoes may range between 152 and 312 million tons per year. Because the actual amount from direct measurements is not known, leading volcanologists today estimate conservatively that annual emissions from volcanoes are 36 million tons per year, in years with no great volcanic eruptions. No matter what figure is used, the basic point remains that the amount of chlorine emitted by Mother Nature through volcanoes dwarfs the amount contained in man-made CFCs.

Furthermore, when there are large volcanic eruptions, much more chlorine is injected directly into the stratosphere. One such famous volcano is Krakatoa, next to the island of Java in the Indian Ocean. In a series of huge eruptions in 1883, Krakatoa sent a shock wave that traveled seven times around the world and covered an area of thousands of square kilometers with ash. Giant waves created by the explosion of the volcano drowned more than 30,000 people in Java. Using very conservative calculations, volcanologists J. Devine, H. Sigurdsson, A.N. Davis, and S. Self (1984) estimated that Krakatoa blew more than 3.6 million tons of chlorine into the atmosphere. Other volcanologists estimate that the amount may have been an order of magnitude greater.

But even the mighty Krakatoa was tiny compared to another volcano in the Sunda volcanic arc in Indonesia, Tambora. When Tambora erupted in the year 1815, some 30 cubic km of the volcano's cone was blown away, injecting enormous amounts of ash and debris directly into the stratosphere. The volcanic cloud reduced the amount of sunlight reaching the surface of the Earth, lowering temperatures. In northern latitudes, the years 1815 and 1816 became known as the years "without a summer," where it

Laurence Hecht

Seawater releases more than 600 million tons of chlorine into the atmosphere every year via evaporation. Most of this chlorine is washed out by precipitation, but large amounts of it still reach the stratosphere.

snowed in the United States in the middle of what was usually the warmest part of the year.

Tambora deposited a layer of ash almost a meter thick as far as 70 km (43 miles) away, and released a *minimum* amount of 211 million tons of chlorine gases into the atmosphere. At present production rates of CFCs, it would take mankind more than 318 years to put as much chlorine into the atmosphere as did the eruption of Tambora. The explosive nature of Tambora and Krakatoa ensured that a significant fraction of the chlorine was carried directly into the stratosphere.

Now, if the ozone-depletion-by-chlorine theory were true, such a catastrophic release of chlorine in 1815 should have wiped out the ozone layer completely, flooding the Earth with so-called cancer-causing ultraviolet rays. Every single man, woman, and child on the Earth should have suffered from skin cancer. Yet, there is no record in the early 19th century of mass extinctions of human, animal, or vegetable life caused by skin cancer or other effects of ultraviolet radiation.

The eruption of El Chichón, a volcano on the Yucatan Peninsula of Mexico, is an indication that large increases in stratospheric chlorine gases do not have a significant effect on the ozone layer. In March and April 1982, there were major eruptions of El Chichón, which injected large amounts of gas and particles into the lower stratosphere. A coherent volcanic cloud was soon established in a zonal band circling the Earth. Several months after the eruption, several aircraft flew through the volcanic cloud measuring the concentration of gases in the stratosphere.

William G. Mankin and M.T. Coffey from the National Center for Atmospheric Research published the results of some of these flights in the Oct. 12, 1984 issue of *Science*. They reported that El Chichón injected more than 40,000 tons of hydrogen chloride directly into the stratosphere, amounting to "about 9 percent of the global stratospheric HCl burden"(p. 171). In the wide band where the volcanic cloud stretched, the amount of stratospheric HCl increased by 40 percent over previous values.

Richard Stolarski and Ralph Cicerone, two of the first proponents of the ozone depletion scare, had originally suggested in 1973 that direct injection of chlorine into the stratosphere by volcanoes could result in substantial ozone destruction. They dismissed this hypothesis in 1974, however, to launch a new scare that chlorine from the Space Shuttle boosters was going to wipe out the ozone layer. In their *Science* article mentioned above, Mankin and Coffin note this turnabout, concluding that their own findings "should lead to a reassessment of the role of volcanoes in stratospheric chlorine chemistry" (p. 172).

On Dec. 14, 1989, Mt. Redoubt in Alaska started erupting and its ash cloud disrupted air traffic, forcing the cancellation of hundreds of flights. A 747 jet aircraft, flying at 25,000 feet with 250 people on board, lost all four engines and nearly crashed before the pilot was able to restart the engines and land safely in Anchorage.

The volcano had a series of explosive eruptions, 24 in total between Dec. 14, 1989 and April 21, 1990. Each eruption created enormous ash clouds, several of which are believed to have penetrated the stratosphere, placing the volcanic debris very high in the atmosphere. Although it was impossible to measure the volcanic gases produced by the volcano until March 20, 1990 (because of the prevailing darkness in Alaska), some estimates were made on the basis of the clouds of ash and sulfuric acid that spread through the United States: The volcano put more than 880,000 tons of sulfur dioxide (SO_2) into the air and an estimated 1 million tons of chlorine gases. Although there have been no explosive eruptions since April 21, 1990, Mt. Redoubt is still active, outgassing 300 to 500 tons of sulfur dioxide every day and approximately 300 to 500 tons of chlorine gases.

These figures on gases from the explosive eruptions are extremely conservative (original estimates were in the range of 2 million tons of sulfur dioxide for the December eruptions alone), but they still make the point that Mother Nature is the worst polluter. The amount of chlorine blown into the air by the volcano is significantly larger than the total amount of chlorine contained in one year's production of CFCs worldwide. As noted before, the gas cloud from the volcano rose all the way to the tropopause in the most explosive eruptions, and it penetrated the stratosphere several times, a number that is uncertain because of the prevailing darkness.

Nevertheless, observations of the eruptions of Mt. Redoubt indicated that a significant amount of volcanic chlorine was injected into the stratosphere in the Northern Hemisphere at the end of 1989 and the beginning of 1990. The volcanic chlorine, therefore, reached the ozone layer, and, therefore, the predictions of the Rowland/Molina theory should have come true in the form of massive destruction of ozone in northern latitudes, accompanied by lethal increases in ultraviolet radiation to residents of—at the least—Canada, New York, and New England in general. There has been no recorded increase in ultraviolet light reaching the ground in those areas. So much for the ozone depletion theory.

All these volcanic eruptions, however, pale in comparison with a more recent series that started on November 17, 1990, with the eruption of Mt. Unzen in Japan, followed in April 1991 by the eruption of Mt. Pinatubo in the Philippines. One of the largest volcanic eruptions of the century, Mt. Pinatubo began as a series of earthquakes, small plumes, and strong explosions, culminating in an eruption June 15–16 that lasted 15 hours. The volcanic cloud reached more than 30 kilometers in altitude, almost to the middle of the stratosphere, creating an enormous ash cloud that was still circling the Earth in 1992.

Another major volcanic eruption, that of Mt. Hudson in Chile from August 12 through 15, 1991, also penetrated the stratosphere. Curiously, the volcanic eruption of Mt. Hudson has barely been mentioned in the news media. Although it occurred in a desolate area of the Andes, it caused massive ecological damage, depositing

U.S. Geological Survey

Mt. Pinatubo in the Philippines in April 1991, one of the largest volcanic eruptions of the century.

more than a cubic kilometer of ash in the Patagonia region of Argentina. Ash deposits 6 inches deep in some places turned almost a third of Argentina into a desert, wiping out crops and killing 1 million sheep.

In a paper published in the Nov. 5, 1991 issue of EOS, Scott Doiron and collaborators estimated that Mt. Hudson outgassed 2.75 million tons of sulfur dioxide into the atmosphere. Their estimates for the eruption of Mt. Pinatubo were almost an order of magnitude greater, 20 million tons of SO_2. Although the figures for chlorine emissions are not yet available, both volcanoes had chlorine- and fluorine-rich magmas, which indicates that tens of millions of tons of chlorine and fluorine were outgassed into the atmosphere and directly injected into the stratosphere by these volcanoes.

Furthermore, the Mt. Hudson eruption created a 2-million-square-kilometer volcanic cloud that traveled directly to Antarctica. Loaded with chlorine and fluorine, the volcanic cloud arrived just in time to circle the polar vortex at the critical moment when the annual ozone hole was being formed; not a word of this volcanic cloud was reported in the press.

As a result of the volcanic eruption of Mt. Pinatubo, there was a flurry of new doomsday predictions, with scientists proposing that the volcano would cause a 15 percent ozone depletion this winter. The villain in this scenario is sulfur dioxide. Guy Brasseur, director of the atmospheric chemistry division of the National Center for Atmospheric Research in Boulder, Colorado, put forward a new ozone depletion theory in 1990. According to this latest entry into the ozone depletion sweepstakes, sulfur dioxide particles from the volcano will undergo a series of complex chemical reactions with other molecules, the final act of which is that stratospheric chlorine reservoir molecules, allegedly from man-made CFCs, will be broken up, freeing up the chlorine atom to gobble up vast quantities of ozone.

Never mentioned by Brasseur is the fact that the same volcanic cloud that lofts the sulfur dioxide molecules to the stratosphere is carrying even greater quantities of chlorine molecules—natural chlorine molecules. Also omitted is the fact that there are volcanic eruptions every year causing great ozone depletions—nothing more than a natural phenomenon—every so often.

Volcanoes and Climate

Now, let's get back to the chlorine that reaches the stratosphere through smaller volcanic eruptions and passively degassing volcanoes. This is an important issue, for as noted above, even in years of no great volcanic eruptions, volcanoes emit more than 36 million tons of chlorine to the atmosphere—4,800 times more chlorine than the theoretical amount of chlorine released every year by the alleged breakdown of CFCs in the stratosphere.

The ozone depletion propagandists dismiss this natural chlorine by arguing that not an ounce of it reaches the stratosphere. The profound scientific issue here is not only how much of this chlorine reaches the stratosphere, but also how volcanoes play a major role in the modulation of climate by changing the "optical" properties of the atmosphere and thus how much sunlight is able to reach the Earth.

It was Benjamin Franklin who first put forward the idea that volcanoes play an important role in climate. In a paper read before the Philosophical Society of Manchester, England, on Dec. 22, 1784, Franklin reported that he had observed a reduction in the intensity of sunlight at the Earth's surface during summer 1783, and he hypothesized that the volcanic eruption of Laki crater in Iceland at the beginning of the summer had created a "dry fog" that was blocking the sunlight. Franklin postulated that the severe winter of 1783–84 in the eastern United States and Western Europe was the result of the reduction of solar intensity, preventing the normal amount of summer heating of the Earth's surface to occur.

Franklin's hypothesis was that the high-altitude "dry fog" was formed from the solid volcanic "dust" ejected by the explosive force of the Icelandic eruption. The volume and injection height of the volcanic "dust" ejected, he said, were directly related to the explosive force of the eruption, the vertical wind structure at the time, and the location of the eruption.

Since Franklin presented this scientific paper in the 18th century, scientists have assumed that only the largest, most violently explosive eruptions would produce a measurable climatic impact. These assumptions have been challenged, however, in the past two decades, as direct samplings of stratospheric aerosols became possible. One of the challengers is J.D. Devine from the Graduate School of Oceanography of the University of Rhode Island. His benchmark paper in 1984 (*Journal of Geophysical Research*) proposed that sulfate aerosols have a greater climatic impact than volcanic "dust." He and coauthors H. Sigurdsson and A.N. Davis found a very close correlation between temperature changes on the surface of the Earth and the amount of sulfur released by a volcanic eruption, a correlation they did not find with other material yielded by the volcanic explosion. This study contained detailed examinations of trace gases estimated to have come from several major volcanic eruptions, including chlorine.

One of the major points of the Devine paper is that it is not necessary to have explosive volcanic eruptions to put all this material, including chlorine, into the stratosphere so that it affects climate. Devine and his collaborators theorized that the thermal structure of the atmosphere over a large lava field may be perturbed by the heat released from the surface of the erupting volcano so that some of the gases released would rise to the stratosphere, a phenomenon they described as "analogous to initiation of free convection of a fluid overlying a flat plate heated from below. . ." (p. 6321).

Another theory was presented in 1984 by Brian Goodman from the Center for Climatic Research at the University of Wisconsin. His doctoral thesis reviews the history of the subject and suggests that low-intensity volcanic activity also has an impact on the climate through "diffuse" sources of volcanic gases. He also suggests that the climate record reflects the existence of harmonic cycles of volcanic activity, influenced by solar and lunar tides.

Before 1970, Goodman states, it was commonly thought that only the solid particles of volcanic dust, called tephra, would reach the stratosphere via violent eruptions. Therefore, the pre-1970 studies looked only at the impact on climate of the largest, most violent eruptions. However, he says,

> ... a more moderately active eruption can produce the same total volume of emissions by compensating for a lower emission rate with a greater duration of activity. In this situation the gaseous emissions are not directly injected into the stratosphere, but are often injected high enough into the upper troposphere that they can remain suspended for several weeks, which will enable some fraction of their original eruption products to be indirectly transported into the stratosphere through one of several stratospheric-tropospheric exchange processes ... [p. 14].

Chlorine and Bromine from the Oceans

Volcanoes may be the most dramatic source of atmospheric gases, but oceans are definitely the greatest source. Extensive studies of the yearly circulation of chlorine and sulfur in nature were carried out in the 1950s and 1960s by Sweden's leading atmospheric scientist, Erik Eriksson. Among Eriksson's many surprising discoveries in examining the transport and exchange of gases in the atmosphere was the fact that 10 times more sulfur passed through the atmosphere yearly than the total amount of industrially released sulfur. Until Eriksson's work, most scientists had assumed all pollution came from man. Eriksson demonstrated that natural sources of pollutants vastly outstripped industrial sources.

In a paper printed in the journal *Tellus* in April 1959, "The Yearly Circulation of Chloride and Sulfur in Nature," Eriksson looked closely at the sources of atmospheric chlorine and calculated that seawater contributes about 600 million tons of chlorine a year to the atmosphere, transported there in 1 billion tons of sea salt particles. (Seawater contains NaCl, sodium chloride, or common table salt. These salt particles enter the air through the evaporation of seawater and through the action of waves, which spray aerosols into the air.)

This means that seawater contributes more than 800 times the amount of chlorine into the atmosphere than is contained in all the CFCs produced worldwide. The comparison with the amount of chlorine allegedly released by the breakup of these CFCs every year is even more startling. There is 80,000 times more chlorine contributed to the atmosphere by seawater than by the alleged breakup of CFCs in the stratosphere. The ozone depletion proponents, of course, claim that not an ounce of this chlorine makes it to the stratosphere because rain washes the chlorine back down to the oceans. They do not bother to explain the origin of the chlorine present in the stratosphere before CFCs were invented. Therefore, there are no references to chlorine from the oceans in any of the ozone doomsday papers, although Eriksson and others provide evidence that indeed vast amounts of chlorine from seawater can and do reach the stratosphere.

Eriksson, for example, set out to determine rigorously not only how much chloride was produced from the oceans, but how long it would remain in the atmosphere, what happened to it, and where it was finally deposited. This was an extremely critical question for Eriksson, since there are large surface salt concentrations in areas of the world that are hundreds of miles away from the ocean. Where does this salt come from? The studies indicate both that large amounts of salt return to the oceans within a short period of time through dry fallout and precipitation, and that large amounts of sea salt could remain in suspension in the atmosphere for long periods of time.

The studies reviewed by Eriksson show that increased turbulence and convective currents over land can carry the oceanic air currents and their salt contents upward into higher altitudes. This is especially the case with warm currrents. Eriksson concludes,

The importance of these findings can hardly be overly stressed as it shows that chloride rich air can be transported at high levels into continents, thus passing over mountain barriers without washout of chloride. Once over continents mixing with colder air will cause precipitation of water and sea salt from such strata. The transport and subsequent accumulation of sea salt in the Bonneville Basin in Utah is thus easier to understand" [p. 397].

Furthermore, tongues of warm air flowing from weather disturbances like thunderstorms are very rich in chlorine. This is extremely important, since thunderstorms, and especially hurricanes, transport enormous amounts of air, chemicals, and water vapor from the surface to the top of the troposphere and can penetrate and inject large amounts of this air directly into the stratosphere.

There are several other conclusive studies demonstrating that vast amounts of chlorine from seawater and other sources do reach the stratosphere. Three scientists from the National Center for Atmospheric Research, A.C. Delany, J.P. Sheldovsky, and W.H. Pollock published a paper in the Dec. 20, 1974 *Journal of Geophysical Research* ("Stratospheric Aerosol: The Contribution from the Troposphere") in which they document the presence of chlorine and bromine in the stratosphere from oceanic sources, noting that there is much more chlorine and bromine in the stratosphere than could be accounted for by direct injection of sea salt particles. Delany et al. write that there is twice as much chlorine as "what would be expected to accompany the sodium as the sea-salt component of the tropospheric aerosol, and bromine is present in an approximately 200-fold excess." They hypothesize that the salt particles, NaCl and NaBr, are broken up in the lower troposphere and then the released gases are carried up to the stratosphere, where they recombine again to produce particulates.

Delany et al. discuss the fact that the chlorine must be from the oceans, but they did not find a clear chemical pathway by which the ions can break up and recombine, and the same is even more true for bromine. One interesting mechanism they propose for the transportation of chlorine and bromine to the stratosphere is "tropopause folding and the smaller scale turbulent transfer between troposphere and stratosphere." It was not until years after this study that other researchers established there were many sources of chlorine in the stratosphere other than sea salt.

Tropospheric-Stratospheric Exchange Processes

That volcanoes and the oceans send vast amounts of chlorine and bromine into the atmosphere is clear. The major scientific issue is, do these gases reach 10 to 16 km up to the stratosphere, and if so, how are they transported?

There are four basic mechanisms to move aerosols and gases from the troposphere to the stratosphere, as discussed in a 1975 paper by Elmar R. Reiter, "Stratospheric-Tropospheric Exchange Processes" (in *Reviews of Geophysics and Space Physics*): (1) seasonal adjustments in the height of the mean tropopause level; (2) organized large-scale motions via meridional circulation; (3) large-scale eddy (vortical) transports, mainly in the jet stream region; and (4) medium- and small-scale eddy transport across the tropopause via hurricanes and other storms that can penetrate the stratosphere and inject into it large amounts of water vapor, warm air, and tropospheric gases.

Here we'll look at the third mechanism, the jet streams. These play a critical role in determining the weather systems throughout the world; yet they are one of the least understood phenomena in the atmosphere. Each hemisphere has three jet streams: one in the subtropics, one in the midlatitudes, and one at the poles— appearing only during certain parts of the year.

The basic point here is that very intense exchanges of air occur in the jet stream, where air from the troposphere is brought up and injected into the stratosphere, and vice versa. Figure 1.3 shows a cross section of a jet stream, with the trajectory of the stratospheric air indicated. The velocity of the wind is greatest in the center, sometimes more than 200 knots, and it is weakest outside the jet stream vortex. Figure 1.4 is a bit more complicated; it shows the cyclonic nature of the jet stream, an anticyclone below it, and how a simultaneous injection of air from the troposphere into the stratosphere occurs and vice versa.

Figure 1.5 is perhaps one of the clearest. It shows a continuous series of measurements of the troposphere-stratosphere interaction carried out by Reinholt Reiter at Zugspitze, the tallest mountain in Germany. (Reinholt Reiter is not related to Elmar Reiter, it should be noted.) Air masses from the troposphere can be seen to enter the stratosphere, several kilometers above the tropopause at the beginning date. Then a stratospheric intrusion breaks downward through the tropopause and from the core of the jet stream there is an injection of ozone-rich stratospheric air down to the surface of the ground. Shortly after this happens, the troposphere bounces back and the opposite occurs: the tropospheric air mass breaks through the tropopause and brings tropospheric air up to an altitude

Figure 1.3
CROSS SECTION OF MASS AIR FLOW FROM THE STRATOSPHERE TO THE TROPOSPHERE NEAR THE JET STREAM

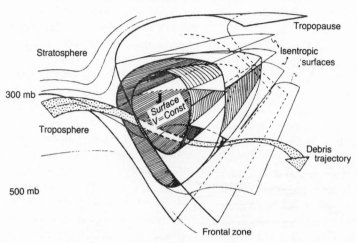

A three-dimensional view of mass air flow from the stratosphere to the troposphere near the jet stream. Isentropic surfaces are indicated by thin lines. Surfaces of constant wind speed, boundaries of the frontal zone, and the tropopause are marked by heavy lines.

Source: Adapted from Elmar R. Reiter, "Stratospheric-Tropospheric Exchange Processes," *Journal of Geophysical Research*, Vol. 13, No. 4 (August 1975).

of 20 km, twice the height of the tropopause. Later, there is another, more intense stratospheric intrusion, where the jet stream again injects ozone-rich, dry stratospheric air to the surface.

(An aside on this issue: There is ozone in the stratosphere, there is ozone at the surface of the Earth, and there is ozone in between. According to standard environmental dogma, ozone in the stratosphere is "good," because it filters out so-called harmful ultraviolet radiation, while ozone in the troposphere is "bad," because it is the main component of urban smog. However, as Figure 1.5 and many other scientific studies demonstrate, stratospheric injections of ozone are very common, and are believed by many scientists to be the main source of tropospheric ozone. As a matter of fact, it is possible for this stratospheric ozone to travel down to the Earth's surface, causing an ozone "smog" alert for which cars in the street are mistakenly blamed.)

Figure 1.4
CYCLONIC NATURE OF THE JET STREAM

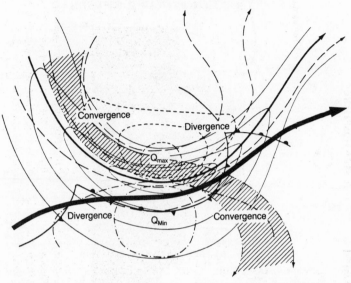

Short dashed lines show lines of vorticity on the cyclonic side of the jet stream, while dash-dot lines show lines of vorticity on the anticylonic side of the jet stream. The shaded band indicates movement of stratospheric air downward into the troposphere; the thicker line crossing it from left to right represents tropospheric air ascending through the jet stream into the stratosphere.

Source: Adapted from Elmar R. Reiter, "Stratospheric-Tropospheric Exchange Processes," *Journal of Geophysical Research*, Vol. 13, No. 4 (August 1975).

Actually, ozone at ground level is getting a bum rap. Although toxic levels have been defined by the U.S. Environmental Protection Agency (EPA), it is beneficial as an ever-present germicide to fight infection of minor cuts and scratches. In fact, hospitals are encouraged to install ozone generators to control the spread of diseases among patients and to help heal wounds from injuries and surgery.

Chlorine from Biomass Burning

Until the late 1970s, the amount of gases contributed to the atmosphere by biomass burning and forest fires was considered

Figure 1.5
EXCHANGE OF AIR BETWEEN THE TROPOSPHERE AND THE STRATOSPHERE

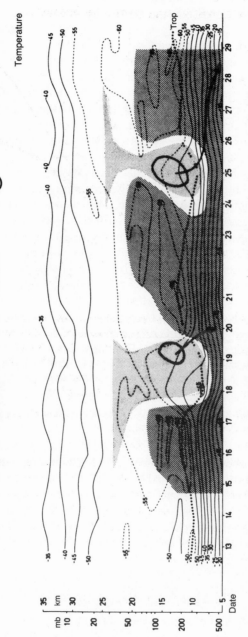

This space-time diagram for October 1980 shows stratospheric air intrusions into the troposphere (light gray areas), while tropospheric air penetrates the stratosphere (dark gray areas). The x-axis represents the time in days; the y-axis altitude. Jet cores and intrusions of stratospheric air (arrows) through the perforated tropopause are indicated. Note how the jet stream forces stratospheric air to penetrate downward and forces tropospheric air upward.

einholt Reiter, "Modification of the Stratospheric Ozone Profile after Acute Solar Events," *Weather and Climate Responses to Solar Variations*, ed. Billy M. McCormac
olo.: Colorado Associated University Press, 1983).

minimal, but careful measurement of forest fires since then indicates that these are a major source of gases for global atmospheric chemistry. Biomass burning, as a matter of fact, may contribute more carbon dioxide to the atmosphere than all of man's industrial activities put together. In a January 1989 interview with one of the authors, scientist Alberto Setzer of Brazil's Institute for Space Studies calculated that more than 540 million tons of CO_2 were released into the atmosphere by the burning of the Amazon rain forest in 1987. American scientist Richard Houghton, in an October 1988 conversation with the authors, put forward a higher figure based on an estimate that includes not only the amount of CO_2 released from the burning of the forest, but also the amount released by the newly exposed soils: 4 billion tons. This means that the burning of the Amazon rain forest alone—which is less than half the world's total forest area burned by man and subjected to primitive agriculture—releases almost as much CO_2 as the *entire* release of CO_2 attributed to industrial activities: 5 billion tons.

Another gas released by biomass burning includes chlorine, the villain of the ozone hole, in the form of methyl chloride (CH_3Cl). According to one of the original papers on the subject ("Biomass Burning As a Source of Atmospheric Gases CO, H_2, N_2O, NO, CH_3Cl, and COS," *Nature,* Nov. 15, 1979, pp. 253–256), 420,000 metric tons of chlorine contained in CH_3Cl were released in 1979 by the burning of biomass. The authors of the paper, Paul Crutzen, Leroy Heidt, Joseph Krasnec, Walter Pollock, and Wolfgang Seiler, caution, however, that their estimates may be very low.

Since this Crutzen et al. study was initially published in *Nature,* the pace of global deforestation, biomass-burning, and slash-and-burn agriculture has increased exponentially—the result of the austerity policies imposed on developing nations and the refusal of international funding agencies like the International Monetary Fund to transfer advanced technologies. Therefore, it is not a surprise that recent, more accurate satellite surveys show global deforestation and burning of the tropical rain forests to be at least 10 times greater than estimated by Crutzen et al. This means that biomass burning is releasing at least 4.2 million tons of chlorine into the atmosphere per year.

This biogenically generated chlorine, according to Crutzen et al., may pose a serious threat to the ozone layer because it breaks up in the stratosphere into the same chlorine atoms produced by the alleged breakup of CFCs. One of the most interesting points of this paper is that it represents another attempt by Paul Crutzen to demonstrate how man is fouling his nest. Crutzen, one of the leading proponents of ozone depletion since the early 1970s, has been rather quiet on this chlorine-from-biomass issue in recent years, perhaps so as not to call attention to the fact that there are sources of chlorine other than man-made CFCs in the stratosphere.

At a March 1990 conference on biomass burning in Williamsburg, Virginia, scientists documented the extent of biomass burning worldwide, not only that caused by man's activities but also that contributed by Mother Nature's forest fires. While world attention was riveted on the great Yellowstone fires in 1989, half a world away, a large area of the Siberian forests and northern China was on fire; millions of acres of forests burned down. In fact, natural forest fires account for most of the biomass burned in the Northern Hemisphere.

The vastness of Mother Nature's forest fires, however, does not lessen the great tragedy occurring in the Third World, as nations are forced to cut and burn their rain forests simply to produce the energy needed for cooking and heating. More than 60 percent of global deforestation is the result of cutting trees for firewood. Most countries in Central Africa obtain more than 90 percent of their energy from biomass burning. Niger, a country that used to export wood, now has to import millions of tons of wood to burn for fuel, because the international lending institutions prohibit it from importing oil and coal, much more efficient energy sources. In fact, most of the global deforestation is the direct result of environmental policies that prevent advanced technologies from reaching the developing sector. Had the attempts to build nuclear power plants in the Third World not been thwarted by the environmentalists, most of the ecological disasters now occurring in those countries would have been averted. If Africans were able to use tractors and fertilizers, for example, they would not have to burn 5,000 square kilometers of savannah every year to plant crops.

To return to the issue of the villainous chlorine: Today, more than 4.2 million tons of chlorine gases are contributed annually by biomass burning, mostly in the tropics. If we take into consideration recent data from the Yellowstone National Park fire and others, reviewed at the March 1990 conference mentioned above, the amount of gases from forest fires in northern latitudes may contribute as much as does biomass burning in underdeveloped nations. Thus, the estimate for the chlorine produced by biomass burning would double to approximately 8.4 million tons. This is at least 11 times more chlorine than that contained in all the world's yearly industrial production of CFCs (750,000 tons) and 1,120 times the amount of chlorine allegedly released from these CFCs.

Even this figure, 8.4 million, however, may be a gross underestimate. Crutzen et al. warn in their 1979 paper that they did not include in their estimates the trace gases that might be released from the heating of the topsoil organic matter or from the "40–80 $\times$ 10^{14} g C [4 to 8 billion tons of carbon] of matter which is exposed to fire but left behind as dead, unburned above-ground biomass" (p. 256). They add that "the topsoil organic matter is especially rich in nutrients and may make important contributions to the cycling of atmospheric trace gases and nutrient elements" (p. 256).

Another point that Crutzen and his coauthors make is the fact that tropical emissions occur in very active atmospheric regions. The two greatest of these dynamic regions are above the Amazon rain forest and the Indonesian Archipelago. These are known as the world's "stratospheric fountains" because of the enormous amount of water vapor and gases transported to the stratosphere by the regions' very violent convective storms.

This is of great importance for the study of chlorine and fluorine transport to the stratosphere, because these tropical rain forests are where the greatest amount of biomass burning takes place, and the most active area of volcanism worldwide is located in the area of the Indonesian stratospheric fountain, providing ready transport for volcanic and biogenic gases to the stratosphere.

The behavior of one of these dynamic regions, the Amazon rain forest, was studied in detail for the first time during a joint atmospheric expedition carried out by NASA and the Brazilian space agency, INPE, in 1985 and 1987. The results of the scientific expedition, called the Atmospheric Boundary Layer Experiment (ABLE), were summarized by Robert J. McNeal, head of NASA's Earth Sciences and Applications Division, in testimony before the U.S. Senate Sept. 20, 1988:

> A mechanism is readily available in the Amazon to transport gases between the planetary boundary layer and the "free" troposphere, where they can enter into large-scale circulation patterns. Deep convective storms of considerable volume are established with great frequency and move essentially continuously around the basin. Such a storm brings material

down from high in the atmosphere, up to and including the stratosphere, which is a source of ozone [and] also raises material from the top of the canopy to the upper troposphere. ... The intensity and frequency of these storms essentially couple the surface and the troposphere vertically in the Amazon region.

These violent storms have enormous power. McNeal says, "Individual convective storms transport 200 megatons of air per hour of which 3 megatons is water vapor releasing 100,000 megawatts of energy into the atmosphere." This is only one thunderstorm; on average, there are 44,000 thunderstorms every day around the world, mostly in the tropics, producing a total of more than 8 million lightning bolts.

The extensive destruction of the tropical rain forest as a result of deliberate policies curtailing technological development is going to have much greater consequences in upsetting the global climate than the imaginary threat from CFCs and other "manmade" pollutants. As McNeal warns in his testimony:

Replacing the forest with wetlands or pasture would likely have large impacts on this enormous furnace with attendant large effects on atmospheric circulation patterns and, therefore, climate. Evapotranspiration and rainfall will decrease. The amount of latent heat transported to temperate and polar latitudes might be reduced, and these regions might experience a cooler climate as a result.

Seaweed vs. CFCs

There are still other vast suppliers of chlorine and bromine to the atmosphere. A 1981 expedition led by atmospheric scientist Hanwant Singh traveled in the Coast Guard Vessel *Polar Sea* from Long Beach, California, to Valparaiso, Chile, taking surface air and water samples. Singh and colleagues discovered that sources in the ocean were emitting vast amounts of methyl halides. Their study, published in the *Journal of Geophysical Research* (April 20, 1983) states, "For the eastern Pacific, mean ocean to air fluxes ... are determined for methyl chloride, methyl bromine, and methyl iodide." These organic chemicals, they explain, "are important carriers of chlorine, bromine, and iodine in the global atmosphere." Singh et al. come to the startling conclusion that their measurements in and over the eastern Pacific demonstrate a dominant oceanic source of methyl halides and that *"this source is large enough to account for virtually the entire tropospheric burden of these species"* (p. 3689 [emphasis added]).

Singh and his collaborators calculated that the oceanic source of these compounds released more than 5 million tons of methyl chloride and 300,000 tons of methyl bromide into the atmosphere. It should be further noted that although their readings of methyl bromide averaged 23 parts per trillion, they measured only 0.7 parts per trillion of halons. Yet, halon is the fire-extinguishing chemical now being banned in the United States *because it contains bromine,* a substance that the ozone depletion faction has labeled as even more evil than chlorine.

In recent years, the ozone depletion scientists have questioned the idea that sources in the oceans could produce such vast amounts of chlorine and bromine—amounts that in and of themselves, without the help of sea salts or volcanoes, are far greater than the amounts of chlorine and bromine in CFCs and halons. The ozone depletion faction argues, without any evidence, that man-made substances must be the source of these methyl halides, not natural sources, because *their models tell them so.* The models, however, don't match the observational data.

The argument between the models and the data was settled conclusively with the July 13, 1990 publication in *Science* of a study conducted by Anne Marie Wuosmaa and Lowell P. Hager from the Department of Biochemistry at the University of Illinois. They demonstrate why marine biomass is responsible for a vast amount of chlorine and bromine injected into the atmosphere, and how it happens.

> The most abundant halohydrocarbon species in the upper atmosphere is methyl chloride, and it is widely believed that biological synthesis is largely responsible for sustaining a global emission rate estimated to be 5 [million] tons of methyl chloride per year. The synthesis of methyl chloride by cultures of wood rot fungi has been well documented, and there have been isolated studies reporting the in vivo synthesis of methyl halides by marine macroalgae and phytoplankton [p. 160].

Wuosmaa and Hager's unique contribution to the field is that they were able to synthesize one of the halohydrocarbons in the laboratory for the first time, methyltribromine. In addition, they discovered an enzyme that can synthesize the methyl halides. The enzyme is present widely in nature, including in fungus, marine red algae, and ice plant. The authors conclude, "Although the production of 5 [million] tons per year represents a prodigious rate of methyl chloride synthesis, this number may be quite understandable in terms of the large terrestrial and marine biomass that can contribute to its formation" (p. 162).

One of the most interesting aspects of the Wuosmaa and Hager paper is their assertion that biogenic sources for the methyl halides may be much greater than previously calculated, both in the oceans and on land. Singh's team took its measurements in the open areas of the Pacific Ocean, but the greatest density of biomass is closer to the coasts. This means production from oceanic biomass may be found to be greater if more extensive measurements are taken.

In addition, Wuosmaa and Hager say that the role of terrestrial plants in the production of methyl halides has been totally ignored. For example, they write,

> The presence of the enzyme [that can synthesize methyl halides] in ice plant, a terrestrial plant which grows in great abundance in the California coastal soils, is an interesting observation that perhaps signals a need for a survey of methyl chloride transferase activity in other succulents that grow in saline-rich environments. Also noteworthy is the fact that ice plant has a wide global distribution [p. 162].

Will Toothpaste Deplete the Ozone Layer?

The other major ingredient of chlorofluorocarbons is fluorine (F), which is also alleged to be an ozone killer and a super "greenhouse gas." (As yet the fluorine in toothpaste and the water supply is not under threat, but such threat is not inconceivable given the sentencing of CFCs.)

Fluorine, like chlorine, is an abundant natural trace gas. The explosion of Tambora in 1815 put up a minimum of 120 million tons of fluorine in the atmosphere. At present production rates, this is the equivalent of 483 years of world fluorine production in CFCs. Furthermore, the amount of fluorine from passively degassing volcanoes may be as high as 6 million tons a year, which is 24 times greater than the world production of fluorine in CFCs, which is approximately 248,600 tons per year. See Table 1.2.

Table 1.2
ATMOSPHERIC SOURCES OF FLUORINE
(tons per year)

Volcanoes	6,000,000
Seawater	44,000
Total natural sources	**6,044,000**
Fluorine in CFCs	248,000
Fluorine theoretically released by the alleged breakup of CFCs	2,480

Three scientists, Robert B. Symonds, William I. Rose, and Mark Reed, published a paper in *Nature* magazine (Aug. 4, 1988) in which they investigate the contribution to the atmosphere of chlorine- and fluorine-bearing gases from volcanoes. After examining the evidence, the authors draw conclusions contrary to the statements of the Ozone Trends Panel. The Ozone Trends Panel "suggest[s] that photolysis of anthropogenic halocarbons in the stratosphere is the only major source of atmospheric HF [hydrogen fluoride]," they write (p. 418). However, "This paper supports other work that naturally degassing volcanoes also emit significant quantities of HF, some of which is directly injected into the stratosphere. Thus, volcanoes should be regarded as a significant source of tropospheric and stratospheric HF."

Another source of fluorine in the atmosphere is the ocean salt sodium fluoride. This source has not been fully investigated, however. Writing in the November 1980 issue of *Reviews of Geophysics and Space Physics,* Richard Cadle says that other than volcanoes "little is known about most other sources of hydrogen fluoride"(p. 749). He estimates that 44,000 tons of fluorine is outgassed annually into the atmosphere from ocean evaporation, but he arrives at this figure by extrapolating from the sodium chloride data. No systematic measurements have been made and the actual rates of fluorine emissions could easily be one to two orders of magnitude greater.

Whodunit?

In all doomsday stories there lurks a villain. According to F. Sherwood Rowland and the rest of the theorists, in the ozone depletion scenario, chlorine from CFCs is the villain. Yet, as we have seen in this chapter, Mother Nature produces orders of magnitude more chlorine from seawater and outgassing volcanoes than CFCs. Why do Rowland and his cothinkers ignore these huge

amounts of chlorine from nature? Is it because they are ignorant—
or because they wish to keep the public ill informed? As we shall
see, this type of lying-by-omission is characteristic of a scientific
scandal with few parallels in history.

References

Steven R. Brantley, ed., 1990. "The Eruption of Redout Volcano, Alaska
December 14, 1989—August 31, 1990." *U.S. Geological Service Circular
1061*.

Richard D. Cadle, 1980. "A Comparison of Volcanic with Other Fluxes of
Atmospheric Trace Gas Constituents," *Reviews of Geophysics and Space
Physics:* Vo. 18, No. 4, pp. 746–752.

T.J. Cassadeval et al., 1990. "Emissions Rates of Sulfur Dioxide and Carbon
Dioxide from Redoubt Volcano, Alaska During the 1989–1990 Erup-
tions." Paper presented at the fall meeting of the American Geophysical
Union, San Francisco, Ca. (Dec. 3–7).

Chapman Conference, 1990. "Global Biomass Burning: Atmospheric, Climatic, and Biospheric Implications." *Proceedings*. Williamsburg, Va.
(March 19–23).

Paul J. Crutzen, Leroy Heidt, Joseph Krasnec, Walter Pollock, and Wolfgang
Seiler, 1979. "Biomass Burning as a Source of Atmospheric Gases CO,
H_2, N_2O, NO, CH_3Cl, and COS," *Nature* (Nov. 15), p. 253–256.

A.C. Delany, J.P. Sheldovsky, and W.H. Pollock, 1974. "Stratospheric Aerosol: the Contribution from the Troposphere," *Journal of Geophysical
Research,* (Dec. 20), Vol. 79, No. 36, pp. 5646–5650.

J.D. Devine, H. Sigurdsson, and A.N. Davis, 1984. "Estimates of Sulfur
and Chlorine Yield to the Atmosphere From Volcanic Eruptions and
Potential Climatic Effects," *Journal of Geophysical Research,* Vol. 89,
No. B7, pp. 6309–6325.

Scott D. Doiron, Gregg J.S. Bluth, Charles C. Schnetzler, Arlin J. Krueger,
and Louis S. Walter, 1991. "Transport of Cerro Hudson SO_2 Clouds,"
EOS (Nov. 5), pp. 489–491.

Erik Eriksson, 1959. "The Yearly Circulation of Chloride and Sulfur in
Nature; Meteorological, Geochemical and Pedalogical Implications,"
Tellus, Vol. 2, No. 4 (Nov.), pp. 375–403.

T.M. Gerlach, H.R. Westrich, and T. J. Casadeval, 1990. "High Sulfur and
Chlorine Magma During the 1989–1990 Eruption of Redoubt Volcano,
Alaska." Paper presented at the fall meeting of the American Geophysi-
cal Union, San Francisco, Ca., (Dec. 3–7).

Brian M. Goodman, 1984. "The Climatic Impact of Volcanic Activity." Ph.D.
thesis, University of Wisconsin, Madison, Wisc.

David A. Johnston, 1980. "Volcanic Contribution of Chlorine to the Stratosphere: More Significant to Ozone Than Previously Estimated?" *Science,*
Vol. 209 (July 25), pp. 491–493.

W.W. Kellogg, et al., 1972. "The Sulfur Cycle," *Science,* Vol. 175 (Feb. 11),
pp. 587–596.

J.P. Kotra, D.L. Finnegan, and W.H. Zoller, 1983. "El Chichón: Composition of Plume Gases and Particles," *Science* (Dec. 2), pp. 1018–1021.

P.R. Kyle, K. Meeker, and D. Finnegan, 1990. "Emission Rates of Sulfur Dioxide, Trace Gases, and Metals from Mt. Erebus, Antarctica," *Geophysical Research Letters,* Vol. 17 (Nov.), pp. 2125–2128.

B.G. Levi, 1988. "Ozone Depletion at the Poles: The Hole Story Emerges," *Physics Today,* Vol. 41 (July), pp. 17–21.

William G. Mankin and M.T. Coffey, 1984. "Increased Stratospheric Hydrogen Chloride in the El Chichón Cloud," *Science,* Vol. 226 (Oct. 12), pp. 170–172.

Lindsay McClelland, David Lescinsky, and Maria Slaboda, 1991. *Bulletin of the Global Volcanism Network,* (May, June, July, and August).

Robert J. McNeal, 1988. "Statement for Senate Hearings on the Implications of Global Climate Change." Presented Sept. 20, pp. 300 ff in hearing record.

R. Monastersky, 1991. "Pinatubo's Impact Spreads Around the Globe," *Science News* (Aug. 31), p. 132.

Nathaniel C. Nash, 1991. "Volcano Ash Is Smothering Vast Areas of Argentina," *The New York Times* (Oct. 21).

Elmar R. Reiter, 1975. "Stratospheric-Tropospheric Exchange Processes," *Reviews of Geophysics and Space Physics,* Vol. 13, No. 4, pp. 459–474.

Reinholt Reiter, 1983. "Modification of the Stratospheric Ozone Profile after Acute Solar Events," *Weather and Climate Responses to Solar Variations,* ed. Billy M. McCormac. Boulder, Colo.: Colorado Associated University Press.

W.I. Rose, R.L. Chuan, and P.R. Kyle, 1985. "Rate of Sulfur Dioxide Emission from Erebus Volcano, Antarctica, December 1983," *Nature,* Vol. 316 (Aug. 22), pp. 710–712.

C.C. Schnetzler, et al., 1990. "Satellite Measurement of Sulfur Dioxide from the Redoubt Eruptions of December, 1989." Paper presented at the fall meeting of the American Geophysical Union, San Francisco, Calif., (Dec. 3–7).

Alberto Setzer, 1989. "The Rain Forest Will Be Gone in 10 to 15 Years" (interview), *21st Century Science and Technology* (Jan.–Feb.), pp. 28–35.

H.B. Singh, L.J. Salas, and R.E. Stiles, 1983. "Methyl Halides in and Over the Eastern Pacific (40°N–32°S)," *Journal of Geophysical Research,* Vol. 88, No. C6 (April 20), pp. 3684–3690.

R.B. Symonds, W.I. Rose, and M.H. Reed, 1988. "Contribution of Cl- and F-bearing Gases to the Atmosphere by Volcanoes," *Nature,* Vol. 334 (Aug. 4), pp. 415–418.

D.C. Woods, R.L. Chuan, and W.I. Rose, 1985. "Halite Particles Injected into the Stratosphere by the 1982 El Chichón Eruption," *Science,* Vol. 230 (Oct. 11), pp. 170–172.

Anna Marie Wuosmaa and Lowell P. Hager, 1990. "Methyl Chloride Transferase: A Carbocation Route for Biosynthesis of Halometabolites," *Science,* Vol. 249 (July 13), pp. 160–162.

2

The Ozone Wars

On July 20, 1969, America's Apollo 11 astronauts landed on the Moon, realizing one of mankind's greatest dreams. As hundreds of millions of people watched and listened, a human being set foot on another body in the solar system. The future seemed bright, and the optimistic expression was born, "If we could land a man on the Moon, then we can certainly. . . ."

The next steps in the U.S. space program were to be the design and construction of an Earth-orbiting space station, and a reusable transportation system to and from orbit. These would be the stepping-stones to a manned mission to Mars, tens of millions of miles from Earth. Technological breakthroughs resulting from the space program had also produced new materials, engine technology, computers, and electronics, which opened up the possibility of developing prototype commercial aircraft able to fly faster than the speed of sound, and, one day, to take off from an airport and fly into outer space.

The supersonic transport envisioned for the next decade would fly as high as the stratosphere at speeds two or three times the speed of sound. Three nations were in the race to engineer and build revolutionary supersonic aircraft: the U.S.S.R., with the Tupolev 144; France, with the Concorde; and the United States, with the

Boeing 2707, known as the SST. Slated for development immediately following the SST, and also under study during the mid-1960s, was a hypersonic plane, which would fly at speeds up to Mach 25. The Air Force Dynasoar (Dynamic Soaring) aircraft, also on the drawing board, would be able to take off from a U.S. airport and land in Tokyo two hours later. It would also be able to fly fast enough to obtain orbital velocity and rendezvous with Earth-orbiting space stations.

Even before the Apollo lunar module set down on the surface of the Moon in 1969, however, an intense fight over the future of the space program and related advanced technologies was taking place on Earth. Virtually as soon as President Kennedy announced the Apollo effort in May 1961, antitechnology think tanks, like London's Tavistock Institute and the Washington, D.C.-based Brookings Institution, were worrying aloud that the space program would ruin their plans for a neo-Malthusian world. By the mid-1960s, Tavistock's Journal *Human Relations* reported that the space program was producing an extraordinary number of "redundant" and "supernumerary" scientists and engineers. "There would soon be two scientists for every man, woman, and dog in the society," one commentator wrote. What worried them most was the climate of technological optimism that had been created.

Their fears were soon allayed. Immediately after the murder of President John F. Kennedy, there was a radical change in national policy. Within days of being sworn in as President, Lyndon B. Johnson dismantled the Kennedy policies that fostered rapid industrial and technological progress, including Kennedy's investment tax credit program. Instead, Johnson initiated the so-called Great Society, under the auspices of which the United States was put on the road to becoming a postindustrial society. The nation's basic industries were dismantled; skilled workers were taken off the production lines. Instead of increasing the wealth of the entire nation through the production of physical goods, the United States would become a service society, with a huge and growing proportion of the population dependent on a welfare state. The end result of this experiment, as we can see today, is a nation that has a standard of living far lower than that of the 1960s.

The space program took a political and financial back seat to the escalating war in Vietnam. Soon, America's premier technological effort was under combined attack from the "budget crisis," which led to drastic cutbacks in government-supported research and development, and from a growing environmentalist movement, bent on destroying high-technology agriculture and industry in America. Development of the commercial SST, the Air Force Dynasoar, the NASA space station, the Space Shuttle, and the Mars mission were crippled by this new Luddite movement.

The fabricated argument that a depletion of the ozone layer would result in a shower of "cancer-causing" ultraviolet rays onto the Earth became one of the most powerful weapons in the antitechnology arsenal. This weapon was wielded without mercy against America's economy through the 1970s and 1980s, in a series of battles that has become known as "The Ozone Wars." The casualties of these wars include the SST project, the Dynasoar, and CFCs, some of the most benign and useful chemicals ever created by man, now banned from use.

The Ozone Wars included mass media propaganda campaigns to convince the public and America's law-makers of the following unproven theories:

(1) That the ozone layer would be depleted by the operation in the stratosphere or mesosphere of supersonic aircraft that exhaust water. When that theory was disproven, nitrogen oxides (NO_x) replaced water as the ozone destroyers.

(2) That the detonation of nuclear devices whose debris clouds can produce or carry NO_x into the stratosphere or mesosphere will deplete the ozone layer.

(3) That the ozone layer would also be depleted by the stimulation of N_2O production by addition of fixed nitrogen to the biosphere whether through nitrogen fertilizers, animal wastes, combustion-produced NO_x, expanded growth of legumes, infection of nonleguminous plants with nitrogen-fixing bacteria, or by green mulching.

(4) That the Space Shuttle would deplete the ozone layer through the release of chlorine from its rocket boosters.

(5) That the ozone layer would be depleted by the atmospheric release of stable chlorine-containing compounds such as chlorocarbons in general and chlorofluorcarbons (CFCs) in particular, which can penetrate the stratosphere before decomposing.

(6) That the ozone layer would be wiped out by the atmospheric release of stable bromine-containing compounds like CH_3Br, now used as a soil fumigant, which can allegedly penetrate the strato-

sphere before decomposing. The same claim was made in regard to brominated chlorocarbons, known as halons, used in fire-fighting equipment.

(7) That the ozone layer would also be depleted by the stimulation of N_2O production by denitrifying bacteria through increased acidity of precipitation from atmospheric release of oxides of sulfur and nitrogen. This theory claimed that the famous "acid rain" in the northern part of the United States would destroy the ozone layer indirectly, through bacteria in the soils.

There is one theory that predicts an *increase* in the thickness of the ozone layer from the release into the atmosphere of stable, infrared-absorbing gases like CO_2, chlorofluorocarbons, and so on, which radiatively cool the stratosphere and thus shift the concentration of O_3. Environmentalists have dropped this theory, however, because it would mean that the so-called greenhouse effect would compensate fully for the "ozone depletion" caused by CFCs.

If several—and in some cases only one—of these claims were true, the atmosphere's ozone layer would have been destroyed *several times over* by today. Yet, as we shall see in the chapters to come, there is no scientific evidence of any ozone depletion.

The SST Controversy

The time is March 1971; the place, congressional hearings on the SST program. Testifying is James McDonald, an atmospheric physicist from the University of Arizona, one of the foremost proponents of the idea that space aliens regularly visit the Earth in UFOs and a passionate opponent of supersonic transport. The ozone depletion theory is about to be unveiled.

Already in circulation were four arguments against the SST project: that the sonic boom from the aircraft would break windows and the eardrums of men and animals; that aircraft noise near the airports would be unbearable; that the SST engine exhaust would pollute the lower atmosphere; and, finally, that climatic changes caused by chemicals in SST exhaust would bring about a new Ice Age. Despite a relentless media campaign pushing these scare stories, Congress had remained committed to building two prototypes of the SST.

Taking the podium to deliver his testimony, McDonald announced a new SST catastrophe theory. His research, he said, had shown that water vapor released by the exhaust of the SST in the

Table 2.1
VARIOUS OZONE DEPLETION SCENARIOS

Ozone depletion theory	% Reduction forecast	Theorist (year)
SST fleet	50	Johnson (1971)
Atmospheric nuclear tests	10	Foley and Ruderman (1973)
Nuclear Summer	70	National Research Council (1975)
CFCs	18	McElroy (1974—to happen by 1990)
Fertilizer	20	McElroy (1976—to happen by 2025)

A sampling of some of the 20-odd ozone depletion scenarios put forward over the past two decades. If all are added up, including the dire predictions that continued use of insecticides and fertilizers would blast away the ozone layer, the depletion should have totaled more than 100 percent by today. Why is there still an ozone layer?

stratosphere would lead to a 4 percent depletion of the ozone layer. And, said McDonald, this ozone layer depletion would result in an additional 40,000 cases of skin cancer in the United States *each year.* The ozone wars had begun.

Congress, however, remained skeptical. Lydia Dotto and Harold Schiff chronicle the events that followed in great detail in their 1978 book, *The Ozone War.* According to Dotto and Schiff: "McDonald came under sharp questioning, but the congressmen seemed more interested in his views on unidentified flying objects, than they were in his concerns about SSTs. McDonald had, in fact, been interested in the UFO problem for some time. He had done a study of UFO data, believed the problem to have been 'scientifically ignored,' and had been a vocal opponent of plans to cancel a UFO observation program" (p. 39). In fact, the last time McDonald had been at a congressional hearing was to testify to his belief that power failures in New York City had been the result of "flying saucers" drawing electricity from power transmission lines.

UFOs or no UFOs, the news media seized on the ozone depletion scare story, which was covered nationally. Within weeks, McDonald was called on to present his theory to the scientific community at a conference in Boulder, Colorado. The meeting, which took place on March 18 and 19, 1971, was sponsored by the Department of Commerce Technical Advisory Board. Its original purpose had been to study the other environmental concerns involving the SST, including the possibility of climate change.

After McDonald's congressional testimony, however, attention was focused on potential depletion of the ozone layer.

The meeting became a battleground. Harriet Hardy, a professor of medicine at Dartmouth Medical School, explained the absurdity of McDonald's skin cancer claims. Arnold Goldberg, chief scientist at Boeing's SST Division, tore apart McDonald's scientific evidence. Goldberg pointed to recent measurements showing that ozone had been increasing in the stratosphere, at the same time that water vapor levels had also been increasing.

This should have been a death blow to McDonald's theory and calculations, but other "atmospheric experts" at the conference argued that observational data were not enough to disprove this hypothetical claim.

It should be noted that all of today's great environmental hoaxes—ozone depletion, global warming, nuclear winter—are based on theoretical models that are contradicted by available, verifiable, observational data. But the so-called scientists who defend these theories claim that all of the *data* are wrong—never the theories. John Swihart, Boeing's chief engineer for the SST project, made exactly this point in a letter to the magazine *Aviation Week & Space Technology* April 12, 1971, shortly after McDonald spoke at the Boulder conference and at the height of the controversy. Swihart documented that all of McDonald's ozone depletion calculations were contradicted by measured ozone and water-vapor data. He charged that the opponents of the SST knew this and "still perpetrated the 'Big Lie' on the American Public. . . . [T]o have publicly condemned the SST by use of a simple model which produced answers in direct conflict with the measured data is, at the least, scientifically dishonest if not treasonous" (p. 60).

That same issue of *Aviation Week* magazine also ran an editorial by Robert Hotz charging that "There was no more cruel and cynical exploitation of the ecological hysteria than the skin cancer issue raised by Dr. James McDonald . . . and Sen. William Proxmire. . . . This was a deliberate big lie concocted knowingly in the face of the measured facts and . . . aimed at creating a wave of false sentiment at a crucial voting time" (p. 11).

McDonald was by no means left on his own to do the dirty work. Also speaking at the Boulder meeting was Harold Johnston, of the University of California at Berkeley. Johnston concocted a new version of the ozone depletion theory on the spot. In Johnston's scheme, it was not water vapor but nitrogen oxides (NO_x), released by the exhaust of the SST operating in the stratosphere, that would deplete the ozone layer.

Dotto and Schiff describe the event:

> Johnston was in an emotional turmoil over the events of the first day. ... [He] stayed up all night to do more detailed calculations on the NO_x problem and prepare a presentation for the next day's sessions. When the delegates assembled on Friday morning, he had photocopies of his handwritten paper ready for them. In it, Johnston calculated that two years of operation by five hundred SSTs would cause global ozone reductions between the "serious" level of 10 percent and the "catastrophic" level of 90 percent, allowing ultraviolet radiation at wavelengths "never before encountered during the evolution of man" to reach the Earth's surface [p. 54].

The participants of the Boulder meeting did not pay much attention to Johnston's theory at the time. Shortly after the meeting was over, however, it became the leading ozone depletion theory. Johnston played a major role in this, by feeding the press with truly frightening stories about how ozone depletion would lead a worldwide epidemic of skin cancer and blindness, induced by overdoses of ultraviolet radiation. According to Johnston, "all animals of the world [except, of course, those that wore protective goggles] would be blinded if they ventured out during the daytime." In a paper published in *Science* magazine in 1971, Johnston predicted ozone depletions up to 50 percent within two years of the advent of the SST.

The SST controversy became fierce. S. Fred Singer, former chief scientist of the U.S. Department of Transportation, wrote in *National Review*, June 30, 1989, summarizing the back and forth arguments of the Ozone War years, under the headline "My Adventures in the Ozone Layer":

> I first got involved in the SST issue in 1970 while serving as a deputy assistant administrator of the EPA. I was asked to take on the additional task of chairing an interagency committee for the Department of Transportation on the environmental effects of the SST. (I had some background in atmospheric physics, having been active in the earliest rocket experiments on the ozone layer, and I even invented the instrument that

Tom Szymecko

S. Fred Singer: "Few outside my special field know about these wild gyrations in the theoretical predictions. But those of us who lived through them have developed a certain humility and affection toward the ozone layer." Singer is shown here speaking at a February 1992 meeting of the American Association for the Advancement of Science in Chicago.

later became the main ozone meter for satellites.) There were many false starts. We knew so little about the upper atmosphere. The ozone problem didn't come up until sometime in 1970, as I recall; and then only in the context of the effects of the water vapor from the burning of the SST's fuel. It was a year later before we came to realize that the main culprit would be, not H_2O, but the small amount of nitrogen oxides (NO_x) created in any combustion process.

The first estimates suggested that some 70 percent of the ozone would be destroyed by an SST fleet; without the ozone shield, "lethal" ultraviolet radiation would stream down to sea level, and an epidemic of skin cancers would sweep the world. This scare campaign led to the cancellation of the SST project. Of course, the two prototypes—all that was authorized—wouldn't have caused any noticeable effect; but the SST opponents had succeeded in confusing the issue. England and France went on to build the Concorde—with no apparent environmental consequences.

Later, it was discovered that there were also natural sources of stratospheric NO_x, and the SST effect soon fell to 10 percent. But then laboratory measurements yielded better data, and by 1978 the effect had actually turned *positive:* SSTs would add to the ozone! It became slightly negative again after 1980, but by then the SST had been forgotten and all attention was concentrated on the effects of CFCs.

Few outside my special field know about these wild gyrations in the theoretical predictions. But those of us who lived through them have developed a certain humility and affection toward the ozone layer. It's a matter of some irony that current theory predicts that aircraft exhaust counteracts the ozone-destroying CFCs. But remember: It's only a theory, and it could change [p. 35].

As Singer recounts, the hysteria created by McDonald and Johnston had done the job it was intended to do: The two prototypes of the SST were killed, and with them the program. Although the United States dumped this program, France and England built the Concorde, while the United States and the Soviet Union went on to build hundreds of aircraft, supersonic bombers, spy planes, and jet fighters—which regularly put as much water vapor and nitrogen oxides into the stratosphere as some of the scenarios proposed by McDonald and Johnston. The ozone layer remains intact.

Hugh Ellsaesser, who works at the federal government's Lawrence Livermore National Laboratory in California and who is one of the world's most respected atmospheric scientists, offered his

Hugh Ellsaesser: *"I can only conclude that during the SST controversy the scientists involved took it upon themselves to act as a priesthood, by suppressing information which the laity could be expected to interpret for itself to arrive at conclusions different from those espoused by the priests."*

comments on the first campaigns of the Ozone Wars in an article published by the journal *Atmospheric Environment* in 1982:

> I can only conclude, that . . . during the SST controversy . . . the scientists involved took it upon themselves to act as a priesthood, by suppressing information which the laity could be expected to interpret for itself to arrive at conclusions different from those espoused by the priests. . . .
>
> For example, data indicating concurrent upward trends in ozone and stratospheric water vapor were not widely circulated as long as water vapor was considered theoretically to be a threat to the ozone layer. Also, all during the Ozone War . . . it was quite clear that the principals did not want the public "to be misled" by being able to equate thinning of the ozone layer with equatorward displacement, presumably because some might begin to wonder what all the fuss was about. Why the discrepancy between theoretical and observational estimates of stratospheric NO_x production rates was ignored remains unclear [p. 204].

Ellsaesser knows what he is talking about. During the past 20 years, he has waged a battle, in most respects a lonely battle, for scientific rigor and truth in the atmospheric sciences. As we show below, in examining the ozone depletion scare stories one by one, Ellsaesser became the leading scientific opponent of the ozone depletion hoax. The hundreds of millions of dollars spent to investigate the validity of many of the climate and atmospheric catastrophe theories of the past 20 years have resulted in data that simply corroborated the criticisms originally made by Ellsaesser in his scientific papers.

From Nuclear Summer to Nuclear Winter

One of the greatest problems faced by the "ozone priesthood" was that the SST threat to the ozone layer could not be rendered realistic by a computer model. The search was on for a real world phenomenon that could be modeled in such a way as to produce computations predicting catastrophic depletion of the ozone layer caused by the injection of nitrogen oxides into the stratosphere.

Nuclear explosions soon became a leading candidate, because the fireballs of nuclear blasts produce enormous amounts of nitrogen oxides that are quickly carried to the stratosphere. The Soviet and U.S. atmospheric tests of nuclear explosives in 1961–1962 were estimated to have injected into the stratosphere an amount of nitrogen oxide comparable to the existing stratospheric inventory of nitrogen oxide or to one year's operation of the projected SST fleet. These tests, therefore, appeared to provide an unparalleled opportunity for validating both the nitrogen oxide catalytic theory and the stratospheric models of ozone depletion.

Scientists H.M. Foley and M.A. Ruderman first suggested the use of nuclear blasts for this purpose in 1973, projecting an ozone reduction of at least 10 percent from the nuclear explosions. They were unable, however, to find any indication of a reduction of stratospheric ozone in the records examined.

As this work proceeded, the perils of nitrogen oxide poisoning of the stratosphere by nuclear explosions soon became a major international issue on its own. Predictions of a "nuclear summer" began to fill major newspapers. This new doomsday scenario predicted that the immediate result of a nuclear war would be the total destruction of the ozone layer, which would allow lethal doses of ultraviolet radiation to reach the Earth. All life on Earth would be wiped out.

This new ozone depletion theory came just at the right time to play a major role in the SALT I negotiations managed by then Secretary of State Henry Kissinger. According to Dotto and Schiff:

> [I]n the fall of 1974, Fred Iklé, director of the U.S. Arms Control and Disarmament Agency, gave several speeches in which he emphasized the hazards to all life on Earth that might result from the ozone depletion caused by nuclear war. His remarks received considerable press coverage. .... Iklé was hopeful that the ozone connection might be a useful bargaining tool in disarmament talks, and he asked the National Academy of Sciences to do a study. ....
>
> The Academy held a five-day workshop in January 1975 and released a report that summer. This report did not consider casualties from the direct hits of belligerent nations, but the aftermath effects of the war, particularly on noncombatant nations. Nor did the study confine itself solely to the ozone question but, as we shall see, the ozone effects were a prominent feature of the report. In fact, the Academy's president, Philip Handler, said that the "principal new point" developed

in the study was that the ozone effect, not dispersion of radiation, would be the major impact on countries not directly involved in the conflict.

The study considered what would happen if ten thousand megatons of nuclear weapons—about half of the then-existing arsenals—were exploded. The conclusion was that the amount of NO_x in the stratosphere would increase by factors of from five to fifty. . . . This in turn would lead to an ozone depletion in the atmosphere over the Northern Hemisphere of from 30 to 70 percent and from 20 to 40 percent in the Southern Hemisphere. The peak effect would occur within a few months of the event, and the atmosphere would take twenty to thirty years to recover. In addition to predicting increases in skin cancer lasting for over forty years, the report said that short-term effects would include "severe sunburn in temperate zones and snow blindness in northern latitudes. . . . For a 70 percent decrease in ozone, severe sunburn involving blistering of the skin would occur in ten minutes" [pp. 302–4].

As was the case concerning all the other hoaxes perpetrated by the ozone depletion theorists, the actual evidence flew in the face of the theory and made a laughingstock of the National Academy of Sciences' report in scientific circles.

According to Hugh Ellsaesser in the 1982 article mentioned previously, there were "many attempts to calculate how much the ozone layer should have been perturbed," and there were extensive "searches of the records from ozone observing stations for evidence that ozone was actually perturbed by these nuclear tests." Ellsaesser listed in a table the investigators and percentage changes in ozone that they calculated or claimed to find in the ozone records.

He reported:

While not too apparent from the table, there have been two dichotomous viewpoints on this issue. On one side are the two U.S. modeling groups represented by Johnston and Chang who have computed global or hemispheric ozone reductions of 1 to 8 percent and then argued that these are supported by, or at least "not inconsistent with" the data. Opposing them have been most of the others who have addressed this issue—

who claim that the observational data do not permit an ozone reduction of more than 1 to 2 percent at most and that this casts doubt on any theoretical results indicating a larger change [p. 199].

In 1973, P. Goldsmith definitively repudiated this "nuclear summer" theory in an article for *Nature* magazine. Goldsmith wrote: "Analysis of the ozone records reveal no detectable changes in the total atmospheric ozone during and after the periods of nuclear weapons testing. Although two models of nitrogen oxide injection [SSTs and nuclear bombs] may not be identical from the meteorological viewpoint, the conclusion that massive injections of nitrogen oxides into the stratosphere do not upset the ozone layer seems inescapable" (p. 551).

The same view was echoed by most leading scientists, among them James K. Angell and J. Korshover, writing in the January 1976 issue of the *Monthly Weather Review*. "If there was a reduction in total ozone following the [nuclear] tests, it is difficult to see how it could exceed 1 to 2 percent. . .," they asserted. "We hereby raise the caution flag, and suggest that perhaps the theoreticians and modelers are in error somewhere along the line, and that at the very least they have overestimated the magnitude of the nuclear (nitric oxide) effect on total ozone" (p. 72).

Since then, it has been discovered that most of the nitric oxide in the atmosphere results not from any of man's activities, but from the *solar wind*. As discussed in greater detail in Chapter 5, the solar wind caries vast amounts of energetic solar protons, which generate nitrogen oxides when they collide with the Earth's atmosphere.

The Space Shuttle Receives the SST Treatment

The ozone warriors continued their offensive in 1973 by setting their sights on the next giant target, the Space Shuttle. NASA projected as the next step in the space program a reusable orbiter that could fly to outer space and fly back and land like conventional aircraft. The orbiter, to debut in 1975, would be used to build space stations in near-Earth orbit, from which the Moon and Mars would be colonized.

In its original conception, the orbiter was to be a craft the size of a medium-range airliner that would ride on top of a booster the size of a 747. Both vehicles would be piloted, and both would be able to fly back to land like conventional aircraft. The booster (a supersonic transport), would have been powered by 12 rocket engines, burning liquid hydrogen and liquid oxygen to lift the paired vehicles from the ground. When the SST had exhausted its fuel, the orbiter would break away and continue into space, using its own hydrogen/oxygen-fueled rockets, leaving the booster to fly back to base under the power of its auxiliary jet engines.

The 1970s, however, saw draconian cuts in the space program budget and also the defeat of the SST. NASA was forced to dangerously compromise the Space Shuttle program and fall back on a cheaper design. The result was the partly reusable Shuttle system that came into operation and is now a familiar sight. This compromise craft is equivalent to the orbiter stage of the original proposal, using its own rocket motors and the fuel from a huge external tank to get into orbit. In order to lift the required weight of fuel off the ground and give the vehicle enough thrust to reach orbit, however, this craft required a pair of additional solid-fuel rockets strapped onto the Shuttle/tank configuration for launch.

As soon as the new configuration of the Shuttle was announced, environmental impact studies began. NASA announced that "no negative environmental effects in the stratosphere are expected as a result of Shuttle operations." But the environmentalists disagreed. They claimed that the hydrogen chloride (HCl) gas emitted by the Shuttle's solid-fuel rockets in the stratosphere could be dangerous. Under pressure, NASA's Marshall Space Flight Center awarded a contract to a team of researchers at the University of Michigan to consider any environmental effects that may have been missed. Why this contract was awarded to the Michigan team is a mystery. The two main researchers were Richard Stolarski and Ralph Cicerone. According to Dotto and Schiff,

> Cicerone and Stolarski were yet another team of "outsiders" when it came to problems of the ozone layer. Like Rowland and Molina, they were not stratospheric chemists. Unlike Rowland and Molina, they were not even chemists. Cicerone had received his degree in electrical engineering and both he and Stolarski, whose training was in physics, had been doing work on the ionosphere, a region of the atmosphere above the stratosphere [p. 125].

But what they lacked in knowledge, Cicerone and Stolarski made up with their enthusiasm to avert a new-found environmental disaster. Dotto and Schiff write:

> Stolarski remembers that they were looking around for something new to do and were attracted by all the activity in the stratosphere. But it was not easy to break into that game ... and "you couldn't just hop into the SST thing. It had been going for too long and we didn't have the credentials." So they decided to take "something that nobody cared about, which was chlorine" [p. 125].

For that purpose, the Shuttle contract was just the ticket. Cicerone and Stolarski's goal became to discover by which mechanism chlorine from the rocket boosters could damage the ozone layer. Assuming 50 flights per year, the boosters would deposit more than 5,000 tons of HCl in the stratosphere annually.

Unbeknownst to Cicerone and Stolarski, another team of researchers was at work assessing the impact of chlorine waste from the Space Shuttle on the stratosphere. The second team, located at Harvard, was led by Mike McElroy and Steven Wofsy. Both teams were working in secret, hoping to accumulate enough scientific evidence to publish a paper and open up the next phase of the Ozone Wars.

Both the University of Michigan and Harvard teams attended an international conference, sponsored by the International Association for Geomagnetism and Aeronomy in Kyoto, Japan, in September 1973. Mike McElroy spoke first, giving an hour-long review of atmospheric photochemistry, dealing primarily with the role of NO_x and barely mentioning chlorine. Richard Stolarski, who immediately followed McElroy, speculated on the role of volcanoes in pumping chlorine into the stratosphere, where the chlorine could deplete the ozone layer. Neither speaker mentioned the Space Shuttle, either from the podium or in the heated debate during the question-and-answer period, when Stolarski came under severe criticism from McElroy.

Most participants at the Kyoto meeting were shocked at the fierceness of McElroy's attacks on Stolarski. They did not know that it was not the role of volcanoes in loading the stratosphere with chlorine that was at stake, but the new ozone depletion theory. If volcanoes and other natural sources could inject chlorine into the stratosphere, then McElroy would never become famous, because the issue would be moot. Stolarski was not being very honest either; the chlorine from the volcanoes was a trial balloon for the chlorine-from-the-Space-Shuttle scare story.

The debate was the beginning of a fierce struggle between the two teams to take credit for the discovery of chlorine as an ozone-busting chemical. According to Dotto and Schiff: "The Kyoto incident left lasting scars. It generated a profound hostility between the Harvard and Michigan groups that would follow them into the fluorocarbon and fertilizer controversies that were to come" (p. 133).

The next battle in the war came in November 1973, when Harold Schiff received a paper from McElroy for inclusion in the proceedings of the conference, to be published in the *Canadian Journal of Chemistry*.. According to Dotto and Schiff, "Schiff was a little taken aback to discover that it was dominated by a discussion of chlorine chemistry, rather than the nitrogen chemistry that had made up the bulk of McElroy's presentation in Kyoto" (p. 131). The paper now stated that the impetus of the studies of chlorine at Harvard was the effect of the Space Shuttle on the ozone layer.

The Harvard team stopped at nothing to prevent its rivals from gaining the credit for the Space Shuttle scare. Harvard team member Steve Wofsy reviewed a paper on chlorine submitted to *Science* magazine by Stolarski and Cicerone and he rejected it for publication.*

When Stolarski and Cicerone, learned of the rejection of their paper, they hit the ceiling. Not only had their paper been rejected, but also they had obtained a preprint of a paper written by Wofsy and McElroy, to be published in the *Canadian Journal of Chemistry,* which dealt with the same subject. (See Cicerone and Stolarski 1974, Wofsy and McElroy 1974.)

During December 1973, several high-level NASA officials decided to look at the scientific evidence concerning chlorine depletion of the ozone layer, and they established the Shuttle Exhaust

---

* Scientific papers are peer reviewed before publication, usually by two or three qualified and *impartial* scientists. Wofsy, as an interested party, should have disqualified himself from such a review. It is unfortunate that the practice of rejecting scientific papers for ideological reasons has become pervasive today. Proponents of the doomsday theories are in control of many scientific journals today and can determine what gets published. Thus, scientific opponents of the global warming and ozone depletion theories find their papers rejected by these journals.

Effects Panel. The panel sponsored a three-day scientific workshop at the Kennedy Space Center in Florida, Jan. 21–24, 1974. Both NASA scientists and outside researchers participated. The workshop concluded that most of the data and calculations were uncertain but that there was a possibility of a 1 or 2 percent ozone depletion from chlorine released by the Shuttle. At least one scientist at the workshop argued that the data indicated that there would be a net increase in ozone.

The workshop's final report stated that Shuttle operations would result in a "small but significant" addition to natural sources of HCl in the stratosphere, adding that the calculations of ozone depletion ranged "from significant to insignificant."

Less than 10 days after the gathering, NASA's Physical Sciences Advisory Committee met at NASA headquarters in Washington, D.C. Harvard's Mike McElroy was the chairman of the public session, where a summary of the workshop was discussed with members of the press in attendance. The results of the discussion came as a complete surprise to many. Without warning, the staff member assigned to present the workshop data announced that a 10 percent depletion of the ozone layer would occur over Florida, in the high-traffic corridor over Cape Canaveral, as a result of Space Shuttle operations. According to Dotto and Schiff, "The numbers were quite a bit higher than any that had been previously discussed [at the workshop] and, according to one participant, very high figures for the climatic effect of this ozone reduction were also suggested. In fact, the numbers came as a bit of a shock to those at the advisory committee meeting, most of whom had never heard of them before" (p. 136). The bomb had been dropped, and McElroy was "bubbling over [about] chlorine chemistry" (p. 136).

In the middle of the afternoon panel, Dotto and Schiff report, "McElroy, who was chairing the session, was handed a note. He looked at it and jumped up from the table. Turning to Donohue, he said: 'Take over' and hurriedly left the room. The note had apparently contained an urgent summons from [NASA Administrator James] Fletcher, and within minutes McElroy was in the administrator's office" (pp. 136–7).

Fletcher was furious, and demanded to know how McElroy could have allowed discussion of this sensitive subject during an open meeting. The question was very serious, said the NASA administrator. It did not matter that the numbers released at the meeting were wild conjectures; the news media could now seize upon them to destroy the Space Shuttle and thus the space program itself.

Dotto and Schiff make the implications of the ozone hoaxsters' actions very clear. "Those events could not have occurred at a worse time," they write. "It was budget time at NASA—open season on the Shuttle as far as critics were concerned. In a few days, Fletcher had to appear before the Senate Space Committee to make his annual defense of the agency's budget of over $3 billion, and there were the usual congressional foes lurking in the wings. The Shuttle was particularly vulnerable. It was NASA's most expensive project; with its annual costs mounting steadily toward the $1 billion mark, it represented nearly a third of the space agency's entire budget" (pp. 137–8).

Fletcher decided to handle the crisis by addressing it publicly. On Feb. 14, during a speech at the National Space Club in Washington, he brought up the speculations concerning the destructive effect of chlorine on the ozone layer and asserted that there were "no data to show that this actually happens." Dotto and Schiff write that "Fletcher, who had not been head of NASA during the SST fight . . . was not worried that the Shuttle would go down the drain like the SST did. But others in NASA indicate that there was certainly concern—and sometimes a fear bordering on panic—that the Shuttle might well share the SST's unhappy fate if it were to be connected, however tenuously, with the ozone controversy" (p. 138).

Many people in NASA believed that McElroy was playing a behind-the-scenes role in creating a media scandal that would bring the chlorine issue forward. After all, McElroy's 1974 paper in the *Canadian Journal of Chemistry* was the first article in the open scientific literature to advertise the hypothesized danger posed to the ozone layer by the Space Shuttle. Whether this is true or not, McElroy suffers from a discernable lack of stature among his peers. Dotto and Schiff report that University of California biochemist Thomas Jukes vividly recalls an incident that tipped him off to McElroy's character: "On a beach at Cape Canaveral in Florida, I saw a red-headed man, sunburned to look like a boiled lobster, applying Novocaine cream to his glowing back. The only unusual circumstance was that the man was Mike McElroy, whose field is the physics and chemistry of planetary atmospheres and who has loudly warned us against the ultraviolet perils of destroying the ozone layer. . . . Surely, he of all people, should have kept his shirt on" (Dotto and Schiff 1978, p. 129).

Hype over Hair Spray

Happily, the Space Shuttle was never added to the casualty lists of the Ozone Wars. It was rescued by still another propaganda campaign asserting ozone depletion, this one centered on an aerosol can of hair spray.

We return to December 1973. The location is the chemistry department at the University of California at Irvine. F. Sherwood Rowland, the chairman of the department, is preparing to depart on a six-month sabbatical to Europe, where he is hoping to get some inspiration to redirect the department's scientific course, as well as some much-needed funding.

Rowland knew one thing that the Harvard and Michigan teams had missed. In 1970, James Lovelock, the official father of the hypothesis that the Earth is a living goddess-creature called Gaia, had built an instrument designed to measure CFCs in the air, at that time one of the most sensitive instruments in the world, capable of measuring parts per million. Lovelock took the instrument on a cruise to Antarctica and back, measuring CFCs all along the way. He discovered the presence of CFCs in Antarctica and concluded that the amounts of CFCs in the air were the equivalent of all the emissions of CFCs from the beginning of their use until then. From this he drew the conclusion that there was no natural process destroying CFCs.

Reviewing Lovelock's findings, Rowland also conjectured that there were no sinks (processes that remove or destroy compounds) for CFCs in the lower atmosphere. He concluded, therefore, that he should direct his attention up to the stratosphere. So Rowland instructed one of his research associates, Mario Molina, to find out what happened to CFCs and the chlorine from CFCs when they reached the stratosphere.

Molina knew nothing about the stratosphere or stratospheric chemistry; his expertise was in chemical lasers. For that matter, neither did Rowland, who had graduated from Ohio Wesleyan University with a major in chemistry and a minor in journalism. Nevertheless, the team proceeded. Shortly before Christmas, after doing calculations on paper for several days, Molina came to

Rowland with a new doomsday theory. He told Rowland that when CFCs rose to the stratosphere, they would be split apart by ultraviolet radiation from the Sun, releasing chlorine. The next step, said Molina, was a catalytic chain reaction in which the chlorine molecules would destroy hundreds of thousands of ozone molecules. His conclusion was that between 20 and 40 percent of the ozone layer would be destroyed.

After poring over the calculations of his assistant, Rowland called Harold Johnston at the University of California at Berkeley. Johnston, as noted above, was already an initiate of the ozone depletion priesthood. Rowland was an ardent follower of Johnston, and had invited Johnston twice to the Irvine campus to give presentations on the SST doomsday theory.

Johnston told Rowland there was nothing new about the chlorine chain doomsday theory, but CFCs as a source of chlorine was something new. Rowland and Molina flew to Berkeley where they met with Johnston. Johnston gave them preprint copies of the scientific papers written by the Michigan and Harvard teams on the threat to the ozone layer from the Space Shuttle and looked over their calculations. At the end of their meeting, Johnston gave Rowland and Molina his blessings and urged them to publish their calculations.

As he left for his sabbatical in Europe, Rowland was on the verge of becoming famous as a result of his assistant's new doomsday theory. He spent his first Sunday in Vienna writing a paper, which he sent to the British journal *Nature*. His next step was to contact the leading proponents of ozone depletion theories in Europe. In Sweden he met with Paul J. Crutzen, who had become—and still is—Europe's grand priest of the ozone depletion theory. Not only did Crutzen give Rowland his blessing, he was actually the first one to unveil the new ozone depletion theory to the press during a speech at the Royal Swedish Academy of Sciences in February 1974, months before Rowland's paper was published in *Nature* magazine.

With the cooperation of the press, however, the CFC scare story remained dormant as long as the Space Shuttle threat to the ozone layer remained a viable scenario to cripple the space program. The CFC story was revived in the fall of 1974, during a meeting of the American Chemical Society in Atlantic City. By then Rowland had joined forces with the University of Michigan team of Ralph Cicerone and Richard Stolarski. In July 1974, the three had met to devise a strategy by which they could break the CFC story in the press. Cicerone and Stolarski submitted an article to *Science* magazine, while Rowland wrote a 150-page paper to support his theory.

In late summer, Dorothy Smith, the news manager of the American Chemical Society, held a press conference for the doomsayers. Rowland used this platform to predict a 50 percent depletion by the year 2050. Cicerone and Stolarski predicted a 10 percent depletion of ozone by 1985 to 1990, previewing a paper that was to appear in *Science* magazine Sept. 27.

Several wire services carried the CFC story, but it did not become major news until it was featured on the front page of *The New York Times* Sept. 26, 1974. The *Times* story, written by Walter Sullivan, was doomsday journalism at its best. In one of the more bizarre turns of the Ozone Wars, however, instead of featuring the work of Rowland, or Cicerone and Stolarski, Sullivan featured a paper written by the competing Harvard team led by McElroy and Wofsy. Sullivan's article created an uproar in the scientific community. The scientific paper featured in *The New York Times* had not even been received by *Science* magazine for publication when the newspaper article came out; the paper was not actually published until February 1975, four months after Sullivan's front-page splash. Furthermore, Dotto and Schiff tell us, "what really caused annoyance and bitterness was the fact that the *Times* story appeared just one day before the Michigan calculations were published in *Science,* in effect scooping them" (p. 23).

Yet one more treacherous sneak attack in the Ozone Wars. By this time, such tactics were very important to the depletion theorists, who believed that a Nobel Prize hung in the balance of the contest.

The paper by McElroy and Wofsy featured in the *Times* had several scenarios for ozone depletion, depending on whether the ban on CFCs was immediate or whether it occurred at a later date. Following the scenario that most closely resembles production rates of CFCs since 1975, the Harvard team's model predicted an 18 percent depletion of the ozone layer by 1990. Whether it be Cicerone's 10 percent or McElroy's 18 percent, the fact is that 1990 has already passed, and there is still no measurable depletion of the ozone layer. Rowland and Molina, the Michigan team, and the Harvard team have been proven to be wrong.

The Bromine Bomb

By 1975, the debate over CFCs was winding down. New grist for the mill was needed. Enter Mike McElroy, fresh with a new theory. During a hearing of the Ad Hoc Federal Interagency Task Force on the Modification of the Stratosphere (IMOS), McElroy announced that bromine, a close relative of chlorine, was more effective at destroying ozone than even CFCs—so effective, in fact, that it would become a lethal weapon if injected over enemy territory. Bromine released into the stratosphere, according to the Harvard scientist, would eat a huge hole in the ozone layer, allowing ultraviolet radiation to reach the Earth to incapacitate enemy troops and civilians and destroy crops. The danger was even greater than this, warned McElroy, because bromine can act like a true "doomsday weapon." The chemical would be carried by winds to the territory of all nations, thus destroying aggressor and victim alike, he said.

McElroy's "bromine bomb" and his demand for an international treaty to ban such weapons made front-page headlines worldwide: "Harvard Professor Warns of ... the Doomsday Weapon.... It's Worse Than the Most Devastating Nuclear Explosion—and Available to All," the *National Inquirer* blared. "A few kilograms of bromine is all that would be needed for a large, devastating effect," McElroy was quoted as saying, "The delivery would be no problem. A small rocket, an aircraft, even a balloon would do. Any country in the world could handle it. And the terrifying thing is that right now, there's nothing to stop them."

McElroy's theory caused a furor in the scientific community. Although the annoyed reaction of some of his colleagues forced McElroy to backtrack somewhat, his claims have had a profound impact to this day. A second element of this theory, which at the time received less attention, was that the agricultural chemical methyl bromide posed a serious hazard to the ozone layer. Methyl bromide is a fumigant used widely against insects and rodents, both in food crops and other commercial crops such as tobacco.

A few months later, Steven Wofsy and his colleagues entered the bromine controversy, claiming an even greater hazard to the ozone layer than methyl bromide. The new entrant to the ozone depletion sweepstakes was brominated chlorocarbons, a chemical closely related to CFCs. These chemicals, more popularly known as halons, are the best fire-fighting chemicals known. The amounts

62

of halons released to the atmosphere are so insignificant, especially when compared to natural sources of bromine, that even the environmentalists did not pay much attention to them until recently. During the March 1990 London Conference, however, halons were included in the CFC ban, without any scientific evidence that they represent a threat to the ozone layer. The next task the environmentalists have set themselves is to ban all fumigants that use methyl bromide, a ban that would devastate modern agriculture.

Ozone Wars in the 1980s

The Ozone Wars continued into the 1980s, with the nuclear summer scenario transforming itself into a nuclear winter scenario.

Ten years to the day after the nuclear summer theory was proposed in 1973, some of the very same scientists who floated that scare story came up with the "nuclear winter" theory. The fact that the same scientists, using the same models, had now come up with the exactly opposite predictions was ignored by the news media. Like the nuclear summer hoax, the nuclear winter theory was proven to be a fraud, and the public did not hear much about it after 1985, until recently. The recent burning of the Kuwaiti oil fields has produced an enormous cloud of dense, black smoke, similar to one that would be produced by use of tactical nuclear weapons. Where is the environmentalists' nuclear winter?

The nuclear winter theory was introduced to the public in November 1983, when ABC-TV network broadcast *The Day After,* a made-for-television "docudrama" that purported to give a realistic picture of the conditions after a nuclear exchange between the United States and the Soviet Union. The broadcast came in a period of great strategic conflict and upheaval. On March 23, 1983, U.S. President Ronald Reagan had announced his controversial Strategic Defense Initiative; six months later, in September, the Soviets had responded to his offer for joint U.S.-U.S.S.R. development of antiballistic missile systems by shooting down the civilian aircraft Korean Air Liner 007.

In addition to *The Day After,* viewers were treated to the performance of television scientist Carl Sagan, who charged in an interview broadcast after the film that it had seriously "downplayed" the consequences of a nuclear war. According to Sagan's calculations, a "small" nuclear exchange equal to about 1,000 megatons of TNT would be enough to obscure the sunlight for weeks and cause temperatures in the Northern Hemisphere to sink 30° Celsius below normal. Even a unilateral preventive nuclear strike, Sagan said, would bring on a "nuclear winter" that would destroy both victor and vanquished. Sagan and colleagues published their calculations in *Science* magazine.

Sagan's nuclear winter hypothesis was summarized by novelist David E. Fisher in his recent book, *Fire and Ice:*

> In the gigantic nuclear-fueled inversion system we are talking about now, the result would be the same. Vertical circulation would be cut down, the particles would remain suspended high up, and sunlight would be cut off from the depths below. TTAPS [Sagan's original report] estimated that continental temperatures would drop by 30°C or 40°C, and that the effect might linger for months or even longer. This gives an average temperature of about −25°C (or −13° Fahrenheit); and temperature drops of only a few degrees, lasting for only a few days or weeks, raise havoc with most food crops. Daylight could not return for many months—a 95 percent cut in sunlight would give days about as bright as normal moonlit nights—and when the sunshine returned it would be to a dead or dying earth.
>
> The world would be subjected to prolonged darkness, abnormally low temperatures, violent windstorms, toxic smog and persistent radioactive fallout [p. 127].

According to this doomsday scenario, transportation systems would break down, as would power grids, food production, medical care, sewage, and sanitation. Government services would be impossible. All over the world there would be starvation, hypothermia, radiation sickness, disease, and death. "Under some circumstances, a number of biologists and ecologists contend, the extinction of many species of organisms—including the human species—is a real possibility," Fisher concludes (p. 127).

Nuclear winter was thus established as a scientific fact. But, was it really?

Absolutely not. Within months, scientists had debunked the computer models used to predict the doomsday scenario. The models used by Carl Sagan, Paul J. Crutzen, and other proponents of the nuclear winter theory had conveniently ignored the existence of the oceans—which cover nearly three-fourths of the surface of the Earth—and their role in the biosphere. Any grade school student who has studied the weather knows that the oceans are the key factor determining weather on Earth. But in Sagan's model there were no thunderstorms, hurricanes, or tornadoes, or ocean-atmosphere interaction. No mention was made of these and other weather phenomena that would clear the troposphere and stratosphere of smoke after a nuclear exchange—as they do after volcanic eruptions.

In an attempt to save their credibility, the nuclear winter theorists claimed that the nuclear summer would follow the nuclear winter. Thus, says Fisher,

> In a nuclear war, enough NO_x could be produced to lower ozone levels by 30 to 50 percent. If anyone was left alive after the slow passage of the nuclear winter, and if those pitiable survivors crawled out of their holes to gaze upward at the emerging sun, their eyes would be burnt out of their heads and their bodies would soon be blackened and burnt to crisps [by ultraviolet radiation] [p. 128].

Soon after Sagan declared the nuclear winter theory, he and others revealed their strategy: The nuclear winter theory "proved" the physical impossibility of winning a nuclear war. Therefore, political and military leaders, East and West, had no choice but to abandon their nuclear strategies, including the purely defensive SDI. It seemed that simple to observers accustomed to Hollywood happy endings.

However, military leaders of both the superpowers had not waited for Sagan's nuclear winter theory. They themselves had raised the question of whether their most powerful weapons might also destroy them. The climatic effects of nuclear explosions— and the immense fires that were expected afterward—had been investigated since the 1960s, netting less spectacular results than those predicted by Sagan et al. Under somewhat more realistic conditions than the worst-case assumptions of Sagan, the climatic effects on the nuclear aggressor are far less devastating than the direct effects of nuclear bombs on the territory attacked.

Thus, the concern of military leaders at the end of the 1970s and the beginning of the 1980s was in exactly the opposite direction of the nuclear winter advocates; namely, that technological development and greater accuracy was increasingly undermining nuclear deterrence.

Gradually, Sagan and his friends conceded that they had exaggerated, and that the substance of the nuclear winter hypothesis could not be maintained. Proponent Stephen H. Schneider came to speak of a "nuclear autumn," and Paul J. Crutzen, whose article in the 1982 *Ambio* had kicked off the nuclear winter thesis, spoke in a newspaper interview not about scientific facts in connection with this theory but rather about political contacts. Physicist Freeman Dyson suggested a simple formula for the nuclear winter: "Good politics, bad physics."

The Latest

We will not entertain the reader with more of the 20-odd ozone depletion theories that have burst onto the scene over the last two decades. One recent development should be noted, however. During the press conference given by NASA scientists Feb. 3, 1992 to announce the discovery of huge amounts of chlorine monoxide above the Arctic (which will allegedly cause an ozone hole there), scientists made another frightening announcement. As the *Washington Post* reported it on the front page the next day:

> In addition, researchers found evidence of reduced concentrations of nitrogen oxides in the lower stratosphere. Nitrogen oxides help preserve ozone by reacting with chlorine and bromine compounds before they can damage the ozone layer. "Our conclusion is that the 'immune system of the atmosphere'—its nitrogen-mediated ability to fight ozone-destroying chemicals—is weaker than we had suspected before," said James G. Anderson of Harvard University, lead scientist for the airborne observations program.

The bottom line is that the same nitrogen oxides declared to be dangerous by opponents of the SST are now described as "the immune system of the atmosphere"!

Also of interest is that in 1992, many veterans of the ozone wars sit in positions of scientific power. Ozone depletion theorist F. Sherwood Rowland, for example, is the president of the American Association for the Advancement of Science (AAAS), publisher

Christopher Sloan

of *Science* magazine. Another ozone warrior, Ralph Cicerone, is president of the American Geophysical Union (AGU), one of the world's leading scientific societies. AGU publishes *EOS,* the *Journal of Geophysical Research,* and *Geophysical Research Letters,* the primary journals through which many scientific theories, especially global warming and ozone depletion, are debated.

Other leading ozone depletion theorists are also in top posts with command power over scientific journals and associations, and, ultimately, public opinion. In this intensely political situation, the doomsday scientific establishment thus decides who is published in the literature and who receives grants—issues that can make or break a career. The doomsday scientists have received a bonanza of research grants, titles, perks, positions, and much more, as a result of the publicity received by their theories. At the same time, the scientists who have had the courage to oppose the doomsday theories in public have had their papers rejected for publication, their grant money discontinued, and in some cases, have even lost their research and teaching positions.

References

J. K. Angell and J. Korshover, 1976. "Global Analysis of Recent Total Ozone Fluctuations," *Monthly Weather Review,* Vol. 104 (Jan.), p. 63.

R.U. Ayres, 1965. "Environmental Effects of Nuclear Weapons." Report HI–518-RR. New York: Hudson Institute.

R.J. Cicerone, R.S. Stolarski, and Stacy Walters, 1974. "Stratospheric Ozone Destruction by Man-made Chlorofluormethanes," *Science,* Vol. 185 (Sept. 27), pp. 1165–1167.

Paul J. Crutzen and J.W. Birks, 1982. "The Atmosphere After a Nuclear War: Twilight at Noon,"*Ambio,* Vol. 11, pp. 114–125.

Paul J. Crutzen, 1989. *Die Welt,* (Oct. 2).

Lydia Dotto and Harold Schiff, 1978. *The Ozone War.* Garden City, New York: Doubleday and Company, Inc.

Hugh W. Ellsaesser, 1978. "Ozone Destruction by Catalysis: Credibility of the Threat," *Atmospheric Environment,* Vol. 12, pp. 1849–1856.

————, 1980. "Man's Effect on Stratospheric Ozone." Proceedings of the Fourth International Symposium on Environmental Biogeochemistry, held in Canberra, Australia, 26 Aug.–Sept. 4, 1979. Canberra: Australian Academy of Science.

————, 1982. "Should We Trust Models or Observations?" *Atmospheric Environment,* Vol. 16, No. 2, pp. 197–205.

David E. Fisher, 1990. *Fire and Ice: The Greenhouse Effect, Ozone Depletion, and Nuclear Winter.* New York: Harper and Row.

H.M. Foley and M.A. Ruderman, 1973. "Stratospheric NO production from Past Nuclear Explosions," *Journal of Geophysical Research*, Vol. 78, p. 4441.

P. Goldsmith et al., 1973. "Nitrogen Oxides, Nuclear Weapon Testing, Concorde and Stratospheric Ozone," *Nature*, Vol. 244 (Aug. 31), pp. 545–551.

Robert Hotz, 1971. "Cancer Charge Refuted," *Aviation Week & Space Technology* (April 12), p. 60.

Harold S. Johnston, 1983. "Global Atmospheric Consequences of Nuclear War." *Science* Vol. 222 (Dec. 23), pp. 1283–1300

Mario J. Molina and F.S. Rowland, 1974. "Stratospheric sink for chlorofluoromethanes: chlorine atomc-atalysed [sic] destruction of ozone," *Nature*, Vol. 249 (June 28), pp. 810–812.

S. Fred Singer, 1988. "Re-analysis of the Nuclear Winter Phenomenon," *Journal of Meteorology and Atmospheric Physics*, Vol. 38, pp. 228–239.

————, 1989. "My Adventures in the Ozone Layer," *National Review* (June 30), pp. 34–38.

Stephen Schneider and Stanley Thompson, 1988. "Stimulating the Climatic Effects of Nuclear War," *Nature*, Vol. 333, pp. 221–227.

Walter Sullivan, 1974. *The New York Times*, Sept. 26.

John Swihart, 1971. "The Ecological Problem," *Aviation Week & Space Technology* (April 12), p. 11.

S.C. Wofsy, M.B. McElroy, 1974. "HO_x, NO_x and ClO_x: Their Role in Atmospheric Photochemistry," *Canadian Journal of Chemistry*, Vol. 52, p. 1582.

Steven C. Wofsy, Michael B. McElroy, and Nien Dak Sze, 1975. "Freon Consumption: Implications for Atmospheric Ozone," *Science*, Vol. 187 (Feb. 14), pp. 535–537.

3

Worldwide Ozone Depletion—Or Doctored Data?

The most recent phase in the ozone wars against CFCs began in March 1988 when NASA's Ozone Trends Panel held a press conference to announce with great fanfare that it had found a 2 to 3 percent global decrease in ozone between 1969 and 1986. The Ozone Trends Panel, created to make an impartial review of all ozone data, threw its weight behind the theory that CFCs are depleting the ozone layer. The panel's conclusion was based on a reanalysis of data from selected ground stations and satellites. Old data that were not considered reliable were either thrown out or subjected to "massage" by statistical mathematical models, apparently to make the data fit the expectations of the panel members.

The panel's final analysis showed a statistical decrease in global ozone of 0.2 percent per year over 17 years. NASA chemist Robert T. Watson, chairman of the panel, pompously told the press that there was no doubt that the panel's analysis "proved" that CFCs were depleting the ozone layer. Even more important, he said, dramatic measures had to be taken to remedy the situation because the data showed even greater depletion than the theory had predicted.

The press was handed an "executive summary" of the panel's report and told that the full report, which was supposed to be released at the press conference, would be made public shortly.

The press conference, with its dire warning, made front-page headlines worldwide. At last, the stories said, here was some real evidence of the much theorized ozone depletion. Various international meetings followed, with the Ozone Trends Panel findings as the centerpiece. Eventually, in June 1990, delegates of 93 nations met in London and decided on a total ban of CFCs. Two major pieces of evidence were presented to the delegates: the Ozone Trends Panel report and data on the so-called ozone hole in Antarctica.

Despite its momentous policy effect, the report of the Ozone Trends Panel was never made available to anyone outside of a select inner circle. In fact, the report did not become public until December 1990, almost *three years later*. From a scientific standpoint, the "executive summary" that had been released to the press and the public March 15, 1988, was worthless. It simply summarized the findings of the panel and did not present any of the data that were utilized to make the analysis. The summary contained none of the scientific details that would enable an independent scientist to judge whether the final analysis of the Ozone Trends Panel was right or wrong.

The mode in which the findings of the panel were announced shocked many in the scientific community. When dealing with scientific issues, especially one of such importance, the usual procedure is to release a report (or an article) on the research to the scientific community for peer review before holding a press conference. The Ozone Trends Panel not only held a press conference on its findings before any peer review, but also never made the full report available for peer review. In fact, for three years, the report remained a well-kept secret, not just to the public but to the scientists who wanted to examine it. Even the final report still lacks the data that would enable an independent scientist to make an objective review.

Even more egregious is that the panel, ostensibly set up to give an impartial judgment on ozone depletion, was comprised of the very same scientists who had authored the ozone-depletion theory! Among the 21 members of the Ozone Trends Panel were Harold Johnston, inventor of the theory that nitrogen oxides from the SST were going to wipe out the ozone layer; Richard Stolarski, inventor of the theory that the chlorine from the Space Shuttle was going to deplete the ozone layer; Richard Turco, one of the inventors of the nuclear winter theory; and, of course, F. Sherwood Rowland,

71

Tom Szymecko

Sherwood Rowland, the eminence grise of the ozone-depletion theory, stands behind a younger proponent of the ozone doomsday theory, Susan Solomon. Both are members of the Ozone Trends Panel. They are shown here at a February 1992 meeting of the American Association for the Advancement of Science in Chicago.

inventor of the CFC depletion theory. Many cothinkers of Rowland, whose names will be familiar from Chapter 2, were either members of the panel or participants in the working groups created by the panel.* The dissenters in the working groups were in the minority; their judgments were ignored and overruled by the panel itself. The biased nature of this arrangement would perhaps be better understood if it were couched in legal terms. Imagine a court case in which the accusers were also the prosecutors, the judge, and the jury. Can one seriously believe that a defendant, in this case CFCs, would stand a chance?

The most astonishing fact about the panel, however, is not its composition but its method. The bulk of the panel's work consisted

---

* The 21 members of the Ozone Trends Panel included Rumen Bojkov, Ivar Isaksen, Harold Johnston, F. Sherwood Rowland, and Richard Stolarski. The panel chairman was Robert T. Watson and the vice chairman was Richard Turco. Participants of the panel's working groups included Ralph J. Cicerone, Michael McElroy, Susan Solomon, and Mario Molina. The one thing these scientists have in common is that their reputations are staked on proving the ozone depletion theory to be correct. In essence, the fox was assigned to guard the hen house.

in examining and "reanalyzing" historical records of ozone data that had been collected at various stations around the world using Dobson spectrophotometers (the main instruments used to measure the thickness of the ozone layer). In other words, members of the panel—most of whom had never operated a Dobson spectrophotometer and who know little of the intricacies of gathering these data—retroactively adjusted the measurements. The effect of this reanalysis is discussed in detail below.

Experimentalists vs. Modelers

Although the major media rarely interview any of the scientists who dissent from the ozone depletion theory, there is actually a deep division in the scientific community today on the ozone issue. One grouping of ozone scientists, who can be classified as experimentalists, spends its time in the field making careful observations of natural phenomena and making hypotheses based on extensive long-term observations. Their hypotheses can then be tested by comparing their predictions to further experimental date.

A second grouping is the ozone modelers, who have gained prominence recently with the advent of supercomputers. The modelers spend their time in the office, selecting data gathered by other scientists to use in making up theoretical models, either on paper or on computers; they then make their hypotheses based on the predictions of their models. These hypotheses almost always center on one aspect or another of how man is destroying Mother Earth. Seldom will the modelers venture out of the confines of their offices to take measurements themselves, unless it is to obtain specific data that will corroborate their model.

The models must be based on a limited number of very simplified assumptions that can be handled by linear equations. They never include all the variables that exist in the real world, owing to lack of knowledge and the limitations of even the most sophisticated computers. Scientists always know that their models are limited tools to work out ideas. But the press and politicians are easily misled to think that the real world conforms to the model— like teenagers who play *Dungeons and Dragons.*

The leading proponents of today's doomsday theories are almost exclusively modelers, and those scientists who oppose them are almost all experimentalists, who base their judgments on observational data, not scenarios and fancy computer printouts.

An anecdote in Sharon Roan's book, *Ozone Crisis,* illustrates this point. In fall 1986, Robert Watson, by then director of NASA's stratospheric research program, put together a team to go to Antarctica and observe the ozone hole firsthand. He asked Susan Solomon, a student of Europe's leading ozone depletion theorist Paul J. Crutzen, to accompany the team. Solomon, who is now looked upon as one of the top experts on the Antarctic ozone hole, accepted reluctantly. As Roan tells the story:

> Solomon created models on computers. She had never done any experimental work, let alone any field work in a place as inhospitable as Antarctica. And, she groaned, she would have to learn how to run Schmeltekopf's instrument [to measure nitrogen dioxide]. But she knew Schmeltekopf had created a solid and well-designed instrument that didn't require a skilled operator [p. 161].

Solomon went to Antarctica together with 12 other scientists and took measurements at McMurdo Station. These are the famous measurements that show concentrations of chlorine oxide in the stratosphere 100 to 1,000 times greater than the concentrations projected by the models. These concentrations were, of course, immediately blamed on CFCs, as proof that man-made CFCs were releasing the chlorine that was destroying ozone and creating the hole. None of these intrepid expeditioners, as pointed out before, bothered to mention to the public that the 33 balloons they launched to take measurements of chlorine concentrations in the air above McMurdo Sound went right through the cloud of volcanic gases from Mt. Erebus, 10 km upwind, which is known to outgas more than 1,000 tons of chlorine every day. Conveniently, the existence of this active volcano is never even mentioned in the accounts of chlorine measured at McMurdo.

Of course, not all the ozone depletion theorists are incompetent at making atmospheric measurements. Some, in fact, are brilliant and have indeed created very sensitive instruments to measure those chemicals they want to find to prove their theories. The problem is that the very specific sensitivity of the instrument does not tell them anything about the other factors that might also be measured. In this single-minded pursuit of the CFC villain, such other factors are overlooked even though they might account for what the scientists are observing. Natural chlorine is one example of a factor that has been overlooked in the search for sources of chlorine in the stratosphere.

Evaluating the Ozone Trends Panel

We will now examine some of the major problem areas in the Ozone Trends Panel's report:

Natural ozone cycles ignored. First, there is the matter of "natural" ozone cycles. As Gordon Dobson demonstrates in *Exploring the Atmosphere* (see Appendix), there are natural cycles that control the abundance of ozone in the stratosphere. Figure 3.1 shows that in northern latitudes the thickness of the ozone layer is almost 500 dobson units in March (spring), dropping down to less than 300 dobson units by October (fall)—a depletion of 40 percent on average each year. In contrast, in the tropics (close to the equator), the thickness of the ozone layer is 220 to 250 dobson units and shows little change on a seasonal basis.

Aside from the seasonal changes that occur to the ozone layer every year in the northern latitudes, there are other, long-term fluctuations of the ozone layer that are not fully understood. The 11-year solar cycle exerts a major influence in a timespan of decades; solar flares play a major role; long-term climate changes, temperature changes, changes in the movement of the jet stream, and planetary waves in the stratosphere all influence the thickness of the ozone layer, both in the short and long term. Because these and many other phenomena influence the thickness of the ozone layer to varying degrees at varying times and can enhance or counteract each other's effects, the long-term cycles of the ozone layer are not uniform.

Subjective measurements. The instruments used to measure the thickness of the ozone layer are very difficult to calibrate; therefore, estimating the actual thickness of the ozone layer involves a great deal of subjective judgment on the part of the observer. For example, the observer has to make a personal assessment by looking at the sky of how much "cloudiness" there is; this subjective judgment is then entered as part of the mathematical formula used to calculate the ozone layer thickness. Such a judg-

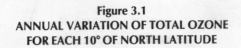

Figure 3.1
ANNUAL VARIATION OF TOTAL OZONE
FOR EACH 10° OF NORTH LATITUDE

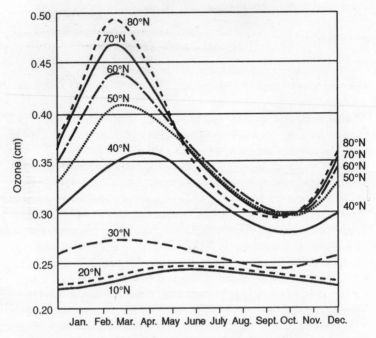

Note the extreme variations in the thickness of the ozone layer especially in northern latitudes between the spring months (March = maximum ozone layer thickness) and the fall (October = minimum layer thickness).

Source: G.M.B. Dobson, *Exploring the Atmosphere,* (London: Oxford University Press, 1968).

ment call has a major effect on the readings, yet how many of these scientists, looking at the same sky, would precisely agree on how much cloudiness there is?

Statistical insignificance. Given all the uncertainty of measurements, the panel's finding of a 3 percent depletion of the ozone layer over 17 years is statistically insignificant. This point is emphasized by S. Fred Singer in an article on the Ozone Trends Panel in the June 30, 1989 issue of the *National Review:*

After subtracting all the natural variations they [the Ozone Trends Panel] could think of—some of them as large as 50 per cent within a few months, at a given station—they extracted a statistical decrease of 0.2 per cent per year over the last 17 years. Making these corrections is very difficult and very technical and very uncertain—especially when the natural variations are a hundred times larger than the alleged steady change.

Prejudicial time frame. The starting date chosen by the Ozone Trends Panel is 1969. Since global ozone data exist as far back as the 1930s, why begin analysis in 1969? As noted by S. Fred Singer and others, 1969–1985 is 17 years, five years short of the two full solar cycles (11 years to a cycle), which is one of the most important long-term influences on ozone. Another important period the panel ignores is the worldwide ozone minimum of 1958–1962. The panel's analysis starts in a year when the thickness of the ozone layer was reaching a maximum, following the very deep minimum observed in 1962. The cycle reached its maximum ozone layer thickness in 1969 to 1970, as shown by Figure 3.2; from there, there was only one way to go—down to a new minimum. Therefore, the choice of years picked by the Ozone Trends Panel in and of itself would ensure that there would be a long-term reduction of the ozone layer thickness observable in the data. If the panel had picked a different 17-year range, say 1962 to 1979, the figures would have shown an actual *increase* in the thickness of the ozone layer!

The year 1962 is particularly important because the ozone layer then was apparently even thinner than it is today. Of course, at that time, CFCs were not in widespread use and therefore could not be blamed as the culprits for the depletion. Hugh Ellsaesser pointed out at the August 1990 conference of the Atmospheric and Geophysical Sciences Division of Lawrence Livermore National Laboratory:

While it is true that the globally averaged depth of the ozone layer appears to have declined some 3 to 5 percent over the past 15 to 20 years, the level today appears to be higher than it was back in 1962. There are at least five papers in the scientific literature reporting 4.3 to 11 percent increases in the depth of the ozone layer from 1962 until the early

77

Figure 3.2
COMPARISON OF SEASONAL VALUES OF SUNSPOT NUMBER
WITH VARIATIONS IN TOTAL GLOBAL OZONE
(1958 through August 1988)

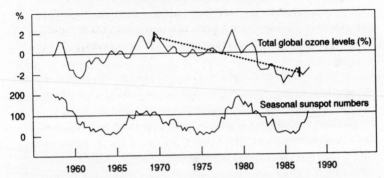

An 11-year and a 22-year cycle in ozone levels, matching the Sun's sunspot cycle, are clearly evident. A large number of sunspots indicates violent disturbance of the Sun's surface, with outbursts of particles and radiation.

Note the 1962 and 1985 ozone minimums. The 1962 ozone minimum is never mentioned by the ozone alarmists. The dotted line from 1969 to 1986 indicates the time frame used by the Ozone Trends Panel to conduct its "analysis" of global ozone data. As Fred Singer and other scientists have noted, the alleged ozone depletion shown is entirely an artifact of the starting and ending dates. Had the ozone trends panel used the same 17-year period (1½ solar cycles), but started in 1962 and ended in 1979, the data would have shown an increase in the thickness of the ozone layer of the same magnitude as the decrease the Ozone Trends Panel reported. In other words, using the same methodology of the Ozone Trends Panel, one can also "prove" that CFCs increase the thickness of the ozone layer!

Source: Adapted from J.K. Angell, "On the Relation Between Atmospheric Ozone and Sunspot Number," *Journal of Climate*, November 1989.

1970s. . . . These numbers are up to twice as large as the declines that have been reported in recent years [pp. 5–6].

Why was there greater global depletion of ozone in 1962 than today? Writing in the November 1986 issue of *Geophysical Research Letters,* Richard Rood, from NASA's Goddard Space Flight Center, points out that dynamical effects may be the causative factor:

The presence of the 1960's minimum suggests that enhanced chlorine levels are not required to produce changes in global total ozone similar in magnitude to the current reduction. Total ozone variations on an annual time scale are largely determined by transport. A quasi-biennial oscillation is obvious. . . . The distinct presence of the minima at stations in the Western Pacific is suggestive of El Niño effects [p. 1246].

A second, and very important point Rood makes is that there was a pronounced decrease in ozone at the North Pole in the four-year period between 1958 and 1962—essentially an Arctic ozone hole. This phenomenon has not been observed recently, and, in fact, those scientists who have been following the behavior of the Arctic ozone layer for the past several decades report that *there has been no overall increase or decrease in the thickness of the ozone layer*. In a January 1990 paper in *Nature* magazine, Norwegian scientists Søren Larsen and Thormod Henriksen analyze Arctic ozone layer data going back to 1935 (Figure 3.3). They conclude:

The data from long-term ozone measurements reveal periods of several years with a negative trend [decrease] and other periods with a positive trend [increase]. The combined results up to 1989 give no evidence for a long-term negative trend of the Arctic ozone layer. . . .
[The data for] Oslo and Tromsø show that the ozone layer over Scandinavia has been above normal (or average) during the past three years. Because of the good correlations with the data from other stations, this conclusion may be valid for the whole Arctic region. . . [p. 124].

Larsen and Henriksen then raise the same critical point raised by Singer: "The figures show the importance of defining the starting point and endpoint when describing trends. The data indicate a positive trend for ozone (in all seasons) in the period 1983–1989 (the past six years). On the other hand, no particular trend can be claimed for the past 10 years"(p. 124). That is, the thickness of the ozone layer undergoes natural fluctuations, and one can show an increase or a decrease, depending on which years are chosen as starting and ending points (Figure 3.3). Nevertheless, they emphasize, *overall, there is no indication of any ozone depletion*.
As Larsen and Henriksen put it: "These data indicate that anthro-

Figure 3.3
NORWEGIAN SCIENTISTS FIND LITTLE
LONG-TERM CHANGE IN ARCTIC OZONE

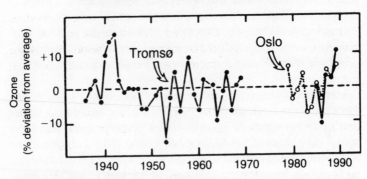

Søren Larsen and Thormod Henriksen at the University of Oslo's Institute of Physics conclude from observational data that gases like CFCs have had a negligible effect on the Arctic ozone layer. "The general balance between formation and destruction of ozone," they write, "has not changed, at least not to an extent that is apparent in the long-term observations."

Shown here are the spring values of ozone for the Norwegian stations at Tromsø at latitude 70°N (filled circles) and Oslo (open circles). The data are the average of measurements in February, March, and April, and correspond to the period when ozone depletion occurs in Antarctica (August, September, and October). These long-term data show that the natural balance between formation and destruction of ozone has not changed in the Arctic.

Source: Adapted from Søren H. Larsen and Thormod Henriksen, "Persistent Arctic Ozone Layer," *Nature* (Jan. 11, 1990).

pogenic gases such as CFCs have, up to the summer of 1989, had a negligible influence on the Arctic ozone layer. The general balance between formation and destruction of ozone has not changed, at least not to an extent that is apparent in the long-term observations" (p. 124).

Larsen, it should be noted, was a student of Gordon Dobson.

Watson Does It Again

Undaunted by the criticism leveled at the 1988 Ozone Trends Panel findings, Robert Watson gave a repeat performance on October 22, 1991, as the cochairman of a panel of the United Nations Environment Program (UNEP). Watson announced new dire ozone results: The ozone layer had been depleted as much as 8 percent, and this time the depletion extended into the summer months. In tandem with Watson, EPA head William Reilly gave a press conference with equally frightening figures. Watson's new data, according to Reilly, would translate into an additional 275,000 skin cancers per year. The UNEP press conference, which had been orchestrated by a paid media consultant, fanned the flames against CFCs once more, giving the opportunity to the environmentalists and those corporations profiting from CFC replacements to call for an even more drastic cutback in CFCs.

The modus operandi was the same as in 1988. Watson presented an "executive summary" of the findings, which he said were contained in a 300-page report to be released by UNEP. However, that report does not exist, according to UNEP spokesmen. The UNEP press conference was followed by congressional hearings Nov. 15, chaired by Senator Albert Gore, in the course of which Gore attempted to link the spread of AIDS and arthritis to ozone depletion!

As S. Fred Singer aptly commented in the *Washington Times* Nov. 20, 1991, "Environmental policy seems once again to be driven by press release rather than by proven scientific data." Singer notes that "with jobs, resources, and half a trillion dollars of equipment in the balance, it seems reasonable to take a long, hard look at the scientific underpinnings of such reports before making major international commitments or taking other steps." He suggested that "with scientists announcing momentous findings by press release, and with the normal peer review process unable to keep pace with policy decisions, perhaps the government, in the public interest, should establish an adversarial kind of proceeding, like a court of inquiry, to allow policy-makers to judge directly the merits of the scientific debate."

How such "momentous" yet unreviewed findings now influence policy worldwide is reflected in a Dec. 13, 1991, newswire by Inter Press Service, reporting on the current debate in the United Nations regarding the monetary allocations of the Global Environment Facility (GEF) and the request of Third World representatives to use the money from the facility to alleviate poverty and hunger in their countries. Inter Press Service reports:

The chair of the UNEP sponsored secretariat for the Scientific and Technical Advisory Panel (STAP), Dr. Robert Watson, said that "global priorities" for GEF fundings had been set by the donor countries. These "priorities" are, controversially, global warming, pollution of international waters, biological diversity, and ozone layer loss—and do not cover poverty issues. The four areas chosen, Watson told members of independent development and environmental groups, had been clearly "politically chosen" by the donors and imposed a definition of global environmental issues.

Watson's Office Does It Once More

Not satisfied with its October performance, Watson's office at NASA went into action again on Feb. 3, 1992. This time, Watson's colleague Michael Kurylo chaired the press conference, announcing that a converted spy plane flying over New England and eastern Canada had detected high levels of chlorine monoxide (ClO) in the stratosphere. This allegedly set the conditions for a huge ozone hole over the Arctic circle, perhaps extending as far south as President Bush's vacation home in Kennebunkport, Maine. Frightening headlines about the "vanishing ozone layer" followed the press conference. *The Washington Post* headlined its Feb. 4 scare story, "Scientists Find Growing Evidence of Ozone-Loss Peril Over Northern Hemisphere." *The New York Times* wrote, "Record Rise in Ozone-Destroying Chemicals Found in North."

Prominently featured in these stories was the threat that "ultraviolet radiation leaking through the ozone layer by the turn of the century could cause 1.6 million additional cases of cataracts and 300,000 additional skin cancers a year worldwide," to quote *The Washington Post*. Senator Albert Gore, made more apoplectic appearances on national television and submitted a bill to the Senate on Feb. 6 calling for a ban on CFCs by 1995. The bill passed unanimously, and on Feb. 11, President Bush seized on this issue to boost his re-election chances by mandating the implementation of this ban.

What are the facts?

First, both the timing of the press conference and the evidence presented were scandalous. This time the ozone depletion theorists gave a press conference before they had even gathered or analyzed the scientific data. According to *The Washington Post,* the

U.S. Geological Survey

The February 1992 ozone scare story was prompted by NASA's finding of increased chlorine in the stratosphere above the Northern Hemisphere. There was little or no mention of the likely culprit—Mt. Pinatubo, which erupted in April 1991. The NASA chlorine discovery coincided with the expected arrival at the poles of the chlorine-rich volcanic cloud from Pinatubo. The cloud is clearly visible in the satellite images released by NASA officials. Here is a view of Mt. Pinatubo erupting.

"indications of ozone depletion by NASA satellite and multiagency airborne instruments [were] so alarming that they decided to release them before completion of the data analysis in late March."

Second, there is no evidence that the ClO molecules detected came from CFCs. The likely culprit is Mt. Pinatubo, which, as noted in Chapter 1, has loaded the stratosphere with millions of tons of chlorine. The NASA discovery of this chlorine coincides with the expected arrival at the poles of the chlorine-rich volcanic cloud from Pinatubo. This cloud is clearly visible in the satellite images released by NASA officials.

Third, the NASA officials noted that the chlorine still has not caused an ozone hole to appear; it simply portends the possibility of a huge ozone hole sometime in the future.

Within days of the Feb. 3, 1992 announcement by NASA that high levels of chlorine monoxide (1.5 parts per billion) were detected over New England and eastern Canada, the Senate passed a bill to speed up the planned phaseout of CFCs and President Bush reversed his previous position and supported a phaseout by 1995.

Finally, the amount of ultraviolet light reaching the Arctic in the middle of the winter is zero, and after the hypothetical ozone depletion it would still be zero. Therefore, there is no need to fit polar bears with sunglasses.

Did 'Reanalysis' Skew the Ozone Data?

Because none of the pronouncements of NASA's ozone doom-sayers during the last two years exists on paper, we must return to the 1988 findings of the Ozone Trends Panel to refute the latest forecast. Even if there were a 3 percent depletion of the ozone layer between 1969 and 1985, as claimed by Watson, this could be accounted for by natural ozone cycles alone. However, the measurements used to arrive at this 3 percent figure are themselves in question. It is necessary to assess whether the massive reanalysis of the ozone data carried out by the Ozone Trends Panel is correct—or simply a self-fulfilling statistical manipulation.

The best guidance in evaluating this case comes from the scientists who spend hours each day making the measurements and analyzing the data from the Dobson spectrophotometers worldwide. Many of these scientists are furious at the Ozone Trends Panel report and other reanalysis of their data records that have been published in the scientific literature by other scientists who have never even set foot in their observatories.

This point was made very clear to one of the authors by Marcel Ackerman, director of the famous Institute d'Aeronomie Spatiale de Belgique (Belgian Institute of Space Aeronomics) in Brussels, during an August 1990 interview: "I don't like the way some people, sitting in their offices, have been transforming data of others [who gathered the data in the field] to prove their point. The reanalyzed data are entirely different from the original values," he said. "These armchair scientists take the data and transform them. They 'correct' the data and then they claim that there is an ozone decrease. But from an ethical point of view this is not at all correct. This is cheating."

The way it should work, Ackerman said, is that

> when there is existing scientific data and someone thinks that those scientists who have collected the data and analyzed it have made mistakes, one gets in touch with those scientists to try to work with them and publish a revision. But this is

not the way some pseudoscientists are proceeding. They take data, they change them, and then they publish as if these were new data.

According to Ackerman, this is what the Ozone Trends Panel did and why their report remained secret for so long. Asked why the Ozone Trends Panel would have to reanalyze the data, if the scientists who had gathered it knew what they were doing in the first place, Ackerman responded: "I don't know. . . . They considered the data not correct for various kinds of reasons. You can always find reasons—especially when you are considering the 2 percent range. You can always claim that here, maybe it was a few percent higher, or it should have been a few percent lower, or something else, and then you come up with new conclusions. . . ."

Those scientists who disagree with the ozone depletion theorists are persecuted, Ackerman said, noting that he had been personally threatened by leading proponents of the ozone depletion theory at three separate scientific conferences:

> If you don't agree with those very strong-speaking people, then you get in trouble, because you are called crazy. They say, "Oh, this guy does not believe in what we say, he is crazy, he should go into a special hospital; you know, a psychiatric hospital." Perhaps if they could they would do as Hitler did, putting those who don't agree into jail. Sometimes it's better not to say you don't agree, because these people are like dictators.

Other scientists have told the authors similar stories. Those who have spoken out openly against the ozone priests at scientific meetings have been threatened, slandered, intimidated, and verbally abused. Many of them have had their scientific papers rejected; some have lost their jobs and research funding. In the scientific community, this is no mere academic debate.

Ackerman also blasted the controlled reporting on the greenhouse effect:

> This is also terrible. In the global warming controversy, they are only publicizing the papers which talk about warming. There are very serious articles where people say we don't see anything, and these are never cited. If you want to have

a paper cited you must be claiming catastrophe. It's very fashionable now.

What should be done about ozone depletion? Ackerman says:

We should wait and see. We don't know enough to know the truth. ... We are not going to live long enough to know the truth, even young people. These geophysical phenomena are very slow, very slow going. It takes a long time.

For me there has not been enough research done and not enough research over a long enough time period. There are many things which are not explained. Even Sherry [Sherwood] Rowland accepts this. He says that these phenomena are so poorly known that you cannot model them. But then in the next sentence he says it will be much worse later than it is now! So in his own article he contradicts himself in the same paragraph! It's incredible.*

In conclusion, Ackerman reiterated his main point:

If you have to manipulate data in order to prove some things, it's dangerous. You may yourself make a mistake. It's very dangerous. And ozone has a great variability. They [the ozone depletion proponents] claim that there is a decrease in ozone at all latitudes, taking data from all over the world, which has been "corrected" and "analyzed." If there is such a kind of manipulation going on, it means that there is no real clear evidence.

Other Challenges from the Scientific Community

Another leading scientist who has challenged the claims of long-term ozone depletion caused by CFCs is C. Desmond Walshaw, a long-time collaborator of Gordon Dobson and former secretary of

---

* The article by Rowland that Ackerman refers to appeared in the January–February 1989 issue of *American Scientist.* Here is the Rowland paragraph in full:
"The problems inherent in modeling of heterogeneous reactions are intractable, especially when the surfaces involved are of uncertain composition and structure, as with the various kinds of stratospheric aerosols and clouds. Any consensus that heterogeneous reactions are significant in any part of the stratosphere almost automatically precludes modeling future ozone depletions, at least in that portion of the stratosphere. The recent discovery that such reactions contribute to extensive

the International Ozone Commission. In an August 1990 interview, Walshaw told one of the authors, "I have grave doubts about some of the inferences apparently being drawn from so-called revisions of Dobson ozone data, especially from tropical regions." Walshaw, who lives in England, pointed out that meteorological changes can have a greater influence on ozone layer thickness than what is being reported as the depletion of ozone caused by CFCs:

> I think the level of change they are claiming is extremely difficult to find in the noise, the natural meteorological changes. If you are talking about temperate latitudes, 3 percent in 20 years is just not significant, because of the noise; the normal year to year variations in ozone are enormous. Furthermore, 3 percent in 20 years is probably smaller than could reliably be detected by the Dobson instrument. If you took all the data that were reasonably reliable, and if all the reliable results showed you a trend of that order, then perhaps you could begin to say it. But who knows this is not a natural trend? There is nothing to tell you that this isn't just a long-term meteorological change.

Commenting on the reliability of the Ozone Trends Panel's claim of 3 percent ozone depletion, Walshaw said:

> I am very skeptical about any data prepared on the midlatitudes in the Northern Hemisphere. ... I am fairly certain that there is no convincing observational evidence for any reduction in midlatitudes. ... One is trying to measure this very small amount of ultraviolet light in the face of an enormous amount of energy in the visible. This is not easy to do at all, and while the Dobson instrument pretty largely succeeds in it, there are still some problems. One has to be aware of this.
>
> The second major problem is that one is asking the instrument to have a long-term stability far higher than any other meteorological instrument. This is an optical photometric stability, which is extremely high. Although methods are being

---

ozone depletion in the Antarctic polar vortex around the spring equinox casts severe doubt on any current predictions of ozone changes—the actual losses in the future are likely to be much more severe [pp. 42–43]."

developed to try to get the stability, it is still not just the present data. It is difficult; problems arise, new problems keep popping up in individual instruments. By no means are these routine measurements.

There are other major problems with the historical ozone data, Walshaw said:

> One of the great difficulties besetting the subject is that the data published by the world data center may or may not have been fully corrected by the observing station. Sometimes it is; sometimes it is not. I think some of the so-called reinterpretation of data is an attempt to correct it, but unless you actually made the observations yourself, and unless you have got very good records of exactly what was done, it is impossible.
>
> A lot of these observations—particularly in the early period, after the International Geophysical Year, in the 1960s and 1970s—vary very much. Some were good and some were not. It is a difficult problem.

Walshaw was most worried about the data used to calculate the alleged 3 percent decrease in ozone:

> The Ozone Trends Panel has got all the data, and they haven't published it and haven't said what they have done with it. The only data publicly available are those produced by the Toronto Data Center [the repository for all global ozone data], and anybody who knows anything about those data, knows that they are extremely inhomogeneous. Some of it has been corrected, and some of it hasn't.

Since the data base and statistical methodology used by the Ozone Trends Panel were kept secret until December 1990, Walshaw, like Ackerman, could not comment on it at the time of his interview. However, Walshaw did know of "revisions" of data that had been carried out by the World Meteorological Organization in Geneva under the scrutiny of Rumen Bojkov, which, he said, for certain had been skewed. The story is that Bojkov was displeased with the data he received from India; it showed no ozone depletion. He then arranged financial support for an Indian scientist to recompile the data. Walshaw, who was personally involved in calibrating the Dobson network of nine measuring stations in India, commented:

> I have studied the so-called adjustments of the one series of ozone observations which I know has been done, the one at Ahmedabad and quite frankly I think the person who did the adjusting does not really fully understand what this is about.
>
> Data taken from Indian stations were reanalyzed, to some extent conscientiously. I mean, the chap looked at all the calibrations, and I think he did his best as he saw it, but I don't believe that his results are reliable, nor do I think anybody else's would be. I think it is inherently extremely difficult to go over this sort of data, which were taken by methods that weren't fully developed at the time.

Specifically, Walshaw pointed to a particular error in the reanalysis, which showed that Bojkov's appointee did not fully understand the consequences of a change made in the standard wavelengths for the observations. In conclusion, Walshaw stressed that the corrections to the Dobson data made by statistical mathematical reanalysis may introduce errors to the data greater than the original errors that the reanalyzers were trying to correct.

Can Models Accurately Predict the Future?

One final issue to address is the accuracy of the computer models that predict the ozone depletion scenario. To put the issue most simply: If climate models can't even accurately predict the weather a week in advance, how can they accurately predict the climate 50 years hence? The news media have done a good job of obfuscating the origins of the doomsday predictions they report as news; most readers are left with the impression that there is hard evidence behind predictions of future ozone depletions or global warming. However, left unexplained is the fact that these predictions are not made on the basis of observed phenomena but on the basis of computer models. With the advent of supercomputers, the models have become even more formidable and convincing to a TV and video-game-oriented audience. In the case of global warming, for example, the computers spit out fancy multicolored maps that show the Earth turning red hot as it warms.

What are these computer models? Let's examine three different kinds used in atmospheric sciences today: weather models, climate models, and chemical models, all quite different from one another.

Weather models are used to forecast the weather in advance. Work on these models has been going on for more than 40 years and, as a result, they are the most powerful and sophisticated. The main data used by weather models are temperature, pressure, wind, humidity, and precipitation readings from thousands of stations around the world. Weather models also receive data from satellites and balloon-borne instruments launched from selected weather stations every six hours to sample conditions in the upper troposphere. Chemistry plays little role in these models.

Weather data are fed continuously into powerful supercomputers, which process the data and make the forecasts. Despite this number-crunching, weather forecasts still have to be interpreted by professional meteorologists—who are themselves often inaccurate.

Climate models are essentially stripped-down versions of weather models, in which the chemistry and physics of the atmosphere play a greater role. For example, a climate model does not rely on daily temperature readings from weather stations. Instead, it makes an enormous grid (with units that are 500 square miles) and assumes that all conditions in this grid are identical—temperature, pressure, wind velocity and direction, humidity and precipitation, to name a few. In climate models, mountains, lakes, rivers, valleys, islands, and so on, do not exist. The inherent absurdities of these assumptions should be clear. Mt. Washington in New Hampshire, the coldest spot in the continental United States, will never have the same climate as Virginia Beach, no matter what the model says.

Such enormous grids are necessary because there are no computers capable of handling the calculations necessary to make more realistic assumptions. These simplifications enable modelers to use a series of complex formulas to examine the changes in climate caused by changes in chemistry (such as the effect of an increase in the concentration of carbon dioxide in the atmosphere) or in physics (such as increased reflection of the Sun's rays caused by increased cloudiness).

The room for error is staggering, to say the least. The change in sign (positive to negative) in just one formula—the one that represents the effect of clouds on climate—could change the result from showing a cataclysmic warming of the Earth in 50 years, to showing an impending ice age. There is a debate now raging in the scientific community over whether clouds warm the Earth by trapping heat, or cool it by reflecting the Sun's rays back into space. Thus, depending on which theory one supports, the same model can make exactly opposite predictions.

If the change in sign of just one formula has such a profound effect in the final forecast, the reader should think about the problems with the several hundred other equations contained in a climate model. In terms of the real world, climate models have proven to be a complete failure. Nonetheless, one can still see them featured in the nightly news as the "certain future" of the Earth if man doesn't stop burning fossil fuels and return to the caves.

Chemical models, like climate models are not nearly as sophisticated as weather models. The physics and especially the dynamics of the atmosphere are ignored in chemical models; the emphasis is on how chemicals interact with each other. Because of the complexity of the chemical reactions involved, however, most chemical models restrict themselves to just a few chemical equations. Great inaccuracies are introduced, because few observational data are used, in contrast to the constant stream of data used for weather models.

Some chemical models incorporate formulas that may be no more than wild guesses at chemical processes. Some include reactions that have never been observed in the stratosphere and exclude some of those that are known to take place, in order to simplify calculations. Furthermore, the models also incorporate what are euphemistically called adjustable parameters. In plain English, this means "fudge factors," which enable dishonest modelers to obtain a desired result and still claim that the computer did it.

Predictions of ozone depletion are based on chemical models. Let's now look at the ranges of ozone depletion predicted by these models:

In 1980, a National Academy of Sciences study predicted an 18 percent ozone decrease based on the standard Rowland/Molina CFCs scenario. This would have meant a 36 percent increase in ultraviolet radiation reaching the surface of the Earth. By 1982, however, a few more atmospheric chemicals had been added to the hypothetical model, and the predicted effect had dropped to a 7 percent decrease. A further refinement of the model in 1984 brought the estimates of ozone depletion caused by CFCs down to 2 percent. The latest claim of the model is that CFCs will deplete the ozone layer by 5 percent over the next 100 years.

So, within a span of four years, the same computer model issued predictions of ozone depletion from CFCs that went from 18 percent, down to 7 percent, down further to 2 percent, and then up to 5 percent. Why such radical changes in the predictions? How could the same model make such wildly divergent predictions? The reason is that after the original, extremely crude theoretical model, the experts started adding other chemicals found in nature to see how this would affect the model predictions. When they added nitrogen oxides, methane, water vapor, and carbon dioxide to the model, suddenly these molecules interacted with the free chlorine. According to the computer simulations, these additions counteracted the effect of CFCs on the ozone layer and the effect went down from 18 percent to 2 percent. Who knows? Additional changes in the model formulas may even show a thickening of the ozone layer, as it did for NO_x emission of SSTs in 1978.

Physicist Freeman Dyson had an interesting insight into why models fail. During his Oct. 11, 1990, Radcliffe Lecture at Oxford University in England, Dyson criticized the doomsday models that predict a greenhouse warming, saying:

...[T]he increased funds have mostly been poured into computer simulations of the global climate rather than into observations of the real world of roots and shoots, trees and termites. I do not blame only the government bureaucrats for the excessive emphasis on computer simulations. We scientists must share the blame. It is much more comfortable for a scientist to run a new computer model in an air-conditioned supercomputer centre, rather than to put on winter clothes and try to keep instruments correctly calibrated outside in the mud and rain. Up to a point, the computer models are useful and necessary. The point at which they are harmful is when they become a substitute for real-world observation. In the 12 years since 1978, the results of computer models have tended to dominate the political discussion of the carbon-dioxide problem. The computer results are simpler and easier for politicians to understand than the vagaries of the real world. The computer results say nothing about non-climatic effects [p. 6].

Basically these modelers operate the same way a witch doctor does. If the magical potion does not have the desired effects, the witch doctor adds a dash of hemlock, a couple of bat wings, perhaps a stretch of lizard skin or a couple of frogs, and so on, until the spell works. Who has a better record, the witch doctors or the modelers? So far, the modelers have yet to verify a prediction.

The final concoction from the modelers' brew is a predicted 5 percent depletion in ozone over the next 100 years; in the short term, this is a figure within the range of the natural oscillations in thickness of the ozone layer. Any more extravagant claim, such as the original 18 percent depletion announced in 1980, would have produced so great a change that every ozone station would have seen a sharp drop in ozone levels already—something that has not happened. By claiming the depletion will occur 100 years hence, the environmentalists can scare the unsuspecting public without having to show any observational data to back up what the model predicts.

Most recently, Robert Watson, Sherwood Rowland, and the other ozone depletion theorists claim that there is evidence of ozone depletion. Their claims, however, refute the models. According to Watson, the recently announced decreases in the ozone layer are *twice as large as predicted in the models*. This means that the model errors are 100 percent; this is of the same magnitude as if the observational data had showed no ozone change whatsoever over the period. If the models are this far off, what basis do the modelers have for claiming that the observed changes are in any way related to what the modelers are predicting?

Is the 100-year prediction reliable? Hardly. As noted in the introduction, at least 192 chemical reactions and 48 photochemical reactions have been observed to occur in the stratosphere. How these very rapid reactions interact with each other is not really known and, therefore, cannot be modeled. In addition, the most recent research shows that the solar wind, cosmic radiation, meteorites, comets, and other atmospheric phenomena interact in a very complex way with the Earth's radiation belt. These, in combination with the dynamics by which the various layers of air in the atmosphere mix, lead to the injection of large amounts of nitrogen oxides into the planet's atmosphere. Electromagnetic vortices at the poles channel high energy particles from the Sun to the surface of the planet, creating such phenomena as aurorae and simultaneously channeling vast quantities of oxygen, ozone, and other chemicals characteristic of life out through the poles up to the magnetotail, millions of miles away from the Earth.

The computer models ignore such phenomena, much as the witch doctor ignores the phenomena of modern medicine in creating his potions.

However, this does not mean that witch doctors do not hold a sway over people. Witch doctors operate on the basis of asserted supernatural powers that they use to scare and awe their devotees. Climate catastrophe doomsayers use same kind of psychological manipulation to achieve their goals. This was revealed by the leading climate modeler in the United States, Stephen Schneider, in an interview with the October 1989 issue of *Discover* magazine. Schneider, who works at the National Center of Atmospheric Research at Boulder, Colorado, said:

> As scientists, we are ethically bound to the scientific method, in effect promising to tell the truth, the whole truth, and nothing but—which means that we must include all the doubts, the caveats, the ifs, ands, and buts. On the other hand, we are not just scientists, but human beings as well. And like most people we'd like to see the world a better place, which in this context translates into our working to reduce the risk of potentially disastrous climatic change. To do that we need to get some broad-based support, to capture the public's imagination. That, of course, entails getting loads of media coverage. So we have to offer up scary scenarios, make simplified, dramatic statements, and make little mention of any doubts we might have. This "double ethical bind" we frequently find ourselves in cannot be solved by any formula. Each of us has to decide what the right balance is between being effective and being honest [p. 47].

In the early 1970s, Schneider was one of the most adamant supporters of the theory that a new Ice Age was about to wipe mankind off the face of the Earth. Today, Schneider is regularly featured in articles and shows warning about impending doom due to global warming. More recently, in February 1992, t' admitted deceiver received the "Scientist of the Year Award" at

Tom Szymecko

Stephen H. Schneider: "We need to get some broad-based support, to capture the public's imagination. That, of course, entails getting loads of media coverage. So we have to offer up scary scenarios, make simplified, dramatic statements, and make little mention of any doubts we may have. . . . Each of us has to decide what is the right balance between being effective and being honest. "

annual meeting of the American Association for the Advancement of Science (AAAS) in Chicago, Ill. The award was presented to Schneider personally by the president of the AAAS, none other than F. Sherwood Rowland.

References

Committee for a Constructive Tomorrow, 1991. "U.N. Fabricates Nightmare on Ozone Street," *Citizen Outlook* (November/December).

G.M.B. Dobson, 1968. *Exploring the Atmosphere.* London: Oxford University Press.

Freeman J. Dyson, 1990. "Carbon Dioxide in the Atmosphere and the Biosphere." Radcliffe Lecture, Green College, Oxford University, Oct. 11.

Hugh Ellsaesser, 1990. "Planet Earth: Are Scientists Undertakers or Caretakers?" Keynote address to National Council of State Garden Clubs, Arlington Hotel, Hot Springs, Ark., Oct. 7.

Søren H. Larsen and Thormod Henriksen, 1990. "Persistent Arctic Ozone Layer," *Nature* (Jan. 11), p. 124.

Sharon L. Roan, 1989. *Ozone Crisis: The 15 Year Evolution of a Sudden Global Emergency.* New York: John Wiley & Sons, Inc.

Richard B. Rood, 1986. "Global Ozone Minima in the Historical Record," *Geophysical Research Letters,* Vol. 13 (November supplement) pp. 1244–47.

F. Sherwood Rowland, 1989. "Chlorofluorocarbons and the Depletion of Stratospheric Ozone," *American Scientist,* Vol. 77, No. 1–2, pp. 36–45.

Kathy Sawyer, 1992. "Ozone-Hole Conditions Spreading; High Concentrations of Key Pollutants Discovered over U.S.," *The Washington Post,* (Feb. 4), p. 1.

S. Fred Singer, 1989. "My Adventures in the Ozone Layer," *National Review* (June 30), pp. 34–38.

———, 1991. "Policy by Press Release," *The Washington Times* (Nov. 20).

4

What Really Happens to CFCs?

The entire edifice known as the Rowland and Molina ozone depletion theory is built on a series of assumptions, and disproving any one of these assumptions will bring the whole house down. This chapter examines—and refutes—two of the most important assumptions: (1) that the only significant sink for CFCs is in the stratosphere through photodissociation by ultraviolet radiation; and (2) that there are no significant sinks that break down CFCs in the lower atmosphere.

Both these assumptions are incorrect. First, the probability that CFCs are being photodissociated in the stratosphere may be as remote as 1 in a billion. Second, there are significant sinks for CFCs in the lower atmosphere, and these can account for the destruction of CFCs that the ozone depletion theory alleges happens only in the stratosphere.

The Rowland/Molina theory holds that CFCs rise unimpeded into the stratosphere, where they are broken up by ultraviolet radiation, freeing a chlorine atom that then goes on a catalytic chain of destruction, destroying hundreds of thousands of ozone molecules.

The first question is: Why doesn't the ultraviolet radiation that reaches the ground break up CFCs at ground level? The reason is that only energetic ultraviolet light shorter than a specific wavelength is capable of doing this, and most of these particular wavelengths are filtered out by the atmosphere as the ultraviolet radiation travels toward Earth. These energetic (shorter wavelength, greater energy) wavelengths, not absorbed by oxygen itself, exist in a narrow channel between 190 and 230 nanometers (nm); only trace amounts of ultraviolet radiation below 290 nm reach the ground.

Therefore, according to the ozone depletion alarmists, in order to meet up with these energetic wavelengths of ultraviolet radiation, CFCs rise to their destruction in the stratosphere. This is the standard explanation the reader will find in the ozone literature and the news media. Missing is the following crucial piece of information: The ultraviolet radiation that reaches the lower stratosphere does not have the wavelengths of ultraviolet at the energy required to break up CFCs. In fact, to encounter any significant concentrations of ultraviolet wavelengths in the 190- to 230-nm band, CFCs have to rise well into the stratosphere, because less than 2 percent of the ultraviolet radiation in the 190- to 230-nm range penetrates below 30 kilometers altitude. Therefore, CFCs have to rise to altitudes of 40 to 45 km (28 miles), almost to the top of the stratosphere (50 km), to find significant concentrations of the ultraviolet wavelengths required for photodissociation.

Do CFCs Rise to the Upper Stratosphere?

CFCs are very heavy molecules, four to eight times heavier than air, depending on which compound is being measured. (For example, a molecule of $CFCl_3$ weighs 137 atomic mass units compared to 32 atomic mass units for O_2.) It is very difficult—but not impossible—for such heavy molecules to rise to the *lower* stratosphere. As noted in Chapter 1, the troposphere is very turbulent, and it diffuses small quantities of gases and particles into the stratosphere, including small amounts of CFCs. The deep convective clouds in the tropics loft air and its contents from the surface boundary layer nearly to, and occasionally through, the tropopause into the stratosphere, in a matter of minutes. However, a large fraction of soluble constituents does not survive this journey, being washed away by the heavy precipitation that accompanies it. This phenomenon is illustrated in Figure 4.1, which depicts measured concentrations of CFCs in the troposphere and the stratosphere.

Figure 4.1
CONCENTRATION OF CFCs AND HALONS IN THE ATMOSPHERE

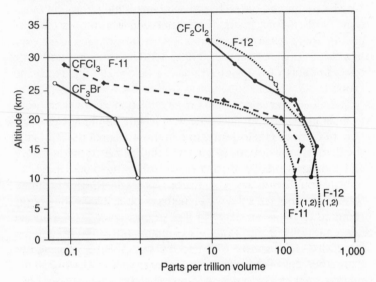

This figure is on a logarithmic scale, with each line on the left representing one-tenth the concentration of the line on the right. Notice how rapidly the concentrations of CFCs and halons decrease after these compounds enter the bottom layer of the stratosphere. The reduction in concentration occurs significantly below the altitude at which high concentrations of ultraviolet radiation (capable of breaking up CFC molecules) are found.

Source: Adapted from R. Fabian, S. A. Borders, and S. Penkett, "Halocarbons in the Stratosphere," *Nature* (Dec. 24, 1981).

Concentrations of CFCs remain fairly steady until they reach the bottom of the stratosphere, as the figure shows. Then, there is a dramatic shift: The concentration of CFCs starts to decrease exponentially. A few kilometers into the stratosphere, the concentrations are only 1/100th of those in the troposphere below.

There are two schools of thought as to what happens to create this abrupt change in the CFC concentrations in these first few kilometers of the stratosphere. One school, the proponents of ozone depletion, pound their chests in triumph and roar that this dropoff is proof positive that CFCs are being broken up by ultraviolet radiation, releasing chlorine atoms to attack the ozone layer. These ozone depletion theorists never mention that no one has observed a single instance of CFCs being split up by ultraviolet radiation in the stratosphere, despite hundreds of millions of dollars in research. The fact is that observing the concentration of CFCs to drop dramatically in the stratosphere is very different from proving that the drop is caused by the breakup of CFCs by ultraviolet radiation.

The second school of thought regarding this issue includes the majority of meteorologists and physical atmospheric scientists, who point to the temperature inversion nature of the stratosphere as the phenomenon that keeps stratospheric populations of CFC molecules small. In an inversion, the temperature lapses are up-side-down; that is, the air becomes warmer with altitude, and the warmer layer of air acts like a lid on the air below.

For those who live in the Los Angeles basin, this is a familiar sight. A recurring low-level inversion is what has caused that basin to be filled with haze during the summer months. During the warm season, a temperature inversion on the top of cool ocean air prevents air from the basin from mixing with warmer tropospheric air above the cool layer. Therefore, the lower, ozone-rich air is bottled up, leading to urban air pollution. This is not a modern development; Indian folklore refers to the basin as "the valley of smoke," and the early Spanish explorers of California reported back to the Spanish crown in the 1600s describing the persistent haze and smoke in the summer.

According to atmospheric scientists, concentrations of CFCs found in the stratosphere are consistent with what one would expect from the effects of an inversion layer on heavy molecules. In this case, the inversion layer is not the top of a cool marine layer, as in the Los Angeles basin, but is the stratosphere itself. Ozone in this layer absorbs solar energy directly, which makes its temperature rise with altitude until—at the base of the still-higher mesosphere—it is almost as warm as the Earth's surface. This inverted temperature lapse rate, many kilometers deep, provides a thermal barrier that resists vertical air movement and thus allows only a few CFC molecules to percolate through the top of the troposphere, or tropopause, into the stratosphere. Furthermore, the CFC molecules that do penetrate the stratosphere and are not decomposed by ultraviolet radiation or mixed at higher levels are recycled back into the troposphere with the normal stratospheric/tropospheric circulation (see Appendix).

This argument is made by Robert W. Pease, professor emeritus in climatology at the University of California at Riverside. Pease asserts that there is a fundamental difference in the thinking of chemists who support the Rowland/Molina hypothesis and the physically oriented atmospheric scientists. He notes, that "CFC–11 and CFC–12 molecules are more than four times the molecular weight of oxygen (O_2)—so heavy that in the laboratory the gases can be poured from one container to another with little spillage. The two CFCs can be used as fire extinguishers because they will form heavy, inert blankets of gas that deprive the flames of oxygen.

"It is precisely because they are so much heavier than air," Pease says, "that it takes considerably more eddy turbulence to support and move the molecules upward through this thermal barrier of the stratospheric inversion than other, lighter air gases. With the insufficient eddy motion that exists at the base of the stratosphere, there will be a gravity bias in the supporting eddy turbulence that causes the net exchange of molecules across the tropopause barrier to have downward bias. This provides a slow-acting sink for stratospheric CFCs that matches the slow rate of percolation upward through the inversion thermal barrier."

The Probability of CFC Destruction

Figure 4.1 makes clear another important point: CFCs do not rise above 40 kilometers, and thus barely reach altitudes where significant energetic ultraviolet radiation(190–230 nm) occurs. As a matter of fact, this ultraviolet radiation has all but disappeared by the time of its penetration to 25 km. The question arises, does enough of the energetic ultraviolet reach the altitudes of CFC occurrence to accomplish much destruction of the molecule?

Pease points out that the same energetic ultraviolet band (190–230 nm) both creates the single atoms of oxygen that form ozone molecules and dissociates the CFC molecules that release chlorine to destroy ozone. The number of CFC molecules relative to oxygen molecules at any point in the stratosphere, therefore, determines the probability of each photoreaction occurring at that point. For measured CFC and oxygen concentrations at 25 km altitude, this probability is roughly that 1 billion ozone molecules will be formed for each CFC molecule that is broken apart. Since the amount of ozone in the stratosphere depends upon a continuous equilibrium between its creation and destruction, the probability of depletion of ozone by CFCs is virtually nonexistent. The destruction of ozone that maintains the equilibrium is by the ultraviolet radiation itself; it is a natural process that has existed as long as there has been an atmosphere that contains oxygen.

In their doomsday models, ozone depletion theorists treat the stratosphere as if it were a homogeneous body. This could not be further from the truth. The stratosphere includes vastly different dynamics and chemistries as a function of altitude and other natural phenomena. How high in the stratosphere CFCs reach is of the greatest importance. Perhaps an illustration familiar to most will help explain this point.

Those who have flown in jet liners have probably noticed a difference between the weather at the airport and the weather on top of the clouds. Cruising altitude for passenger jets is 25,000 to 35,000 feet—the top of the troposphere in midlatitudes. A thunderstorm, with tornadoes, may be raging near the ground, but passengers in the jet plane see only the cloud tops and may experience a little turbulence.

Similarly, very different weather can occur over the same spot at different altitudes in the stratosphere. What goes on in the middle and top of the stratosphere is very different from what goes on at the bottom of the stratosphere, and this applies to weather, chemistry, photochemistry, atmospheric electricity, and so on. The issue of atmospheric electricity is particularly important. Atmospheric electricity, one of the least understood natural phenomena, plays a critical role in determining the path and speed of chemical reactions in the atmosphere. Yet, this variable is completely ignored in all the models used by the ozone scaremongers.

Enormous amounts of electricity are involved here. For example, one major thunderstorm, the kind that spawns tornadoes, may produce more electricity in a few hours than the entire amount of electricity generated by the U.S. electric grid in one year. Ignoring atmospheric electricity is not a trivial matter.

Disappearing CFCs

In great dispute in the scientific community today is the question of the amount of CFCs that disappear every year. The debate is framed in terms of the expected lifetime of CFCs in the atmosphere, a lifetime being the time for 68 percent of the CFCs in the atmosphere to be destroyed. This number is arrived at by *estimating* the total amount of CFCs produced since they were first manufactured in 1930, and also *estimating* how many of these CFCs have actually been released. Scientists make measurements of CFCs at different stations around the world, and from these measurements they *estimate* the total amount of CFCs in the atmosphere today. The theoretical amount that should be in the atmosphere is then subtracted from the amount that is *estimated* to be there, some complex statistical calculations are carried out, and the estimated lifetime is arrived at.

These estimated lifetimes range from 20 years to 1,000 years and are very controversial. There are two reasons for the controversy: first, the concentrations of CFCs in the atmosphere are infinitesimal; second, the discrepancies in the estimates of CFC production and release are enormous.

The most common figures used are lifetimes of 75 and 120 years for CFC–11 and CFC–12, the most abundant CFCs. These figures are still controversial, but the average here is useful in illustrating our point. Yearly production of CFCs is 1,100,000 tons. If an average lifetime of 100 years is assumed for CFCs, that means that 68 percent of the present CFCs in the atmosphere will be removed in the next 100 years. To simplify these calculations, let's say that 1 percent of 1.1 million tons of CFCs (11,000 tons) is destroyed each year. Those 11,000 tons of CFCs will release 7,500 tons of chlorine. (This is a minuscule amount, as noted in Chapter 1, compared to nature, which spews more than 650 million tons of chlorine into the atmosphere every year.)

Now, since approximately only 1 percent of the CFCs in the atmosphere disappears every year, the Rowland/Molina theory assumes that there are no other significant sinks for CFCs on Earth and that CFCs are broken up in the stratosphere by ultraviolet radiation, nothing else. If there were any natural sinks for CFCs elsewhere, then the Rowland/Molina theory would fall apart—and that is exactly what we will show: There are indeed significant sinks for CFCs in the troposphere, sinks that can account for every ounce of CFCs that has "disappeared." Considering how many lives are at stake in the CFC issue, it is astonishing that the ozone depletion theorists did not even investigate this possibility.

There indeed appear to be significant sinks for CFCs in the troposphere. Table 4.1 lists those sinks that have been reported in the scientific literature. Some of these are sinks where CFCs are destroyed; some are sinks where they are deposited.

Table 4.1
ATMOSPHERIC SINKS OF CFCS

Soils (deposition)
Soil bacteria (destruction)
Biomass (captured by plant lipoproteins)
Oceans (deposition)
Ocean biota (destruction)
Desert sands (destruction)

Environmental Microbiology

One of the frontiers of science today is little known but of great importance to our future—environmental microbiology. Twenty years ago, it was a commonly held belief that man-made synthetic compounds could not degrade and be broken up in the environment. Today, scientists know that the opposite is the case. For every compound created by man, there are bacteria or bacterial enzymes that can destroy it and turn it into either food or energy. The study of these specialized bacteria is the field of environmental microbiology.

Microbes that eat oil spills are old news. Newer items are the microbes that capture and concentrate radioactive waste and the most foul toxic wastes ever produced by man. And there are even microbes that devour farm chemicals. Farmers have been surprised when the herbicides and pesticides they have used for decades disappear in less than a week, instead of lasting up to three years in the fields. Scientists at Battelle Pacific Northwest Laboratories discovered after extensive studies in farm fields in Washington state that colonies of bacteria have evolved over the years that consider those herbicides and pesticides a source of food—and act accordingly.

At present, a team of scientists at Battelle is looking for microbes that will destroy specific toxic pollutants. They are drilling holes thousands of feet deep to find new microorganisms that can be used at the Earth's surface to destroy toxic waste. Worldwide, thousands of microbiologists are involved in this fascinating and intricate pursuit. The stakes are very high. Groundwater pollution is a real problem, and in the United States alone it is expected that cleaning up existing toxic waste sites and leaking underground storage tanks is going to cost more than $500 billion in the next few decades. The cost could be reduced to a fraction of that amount with the research now being conducted on *bioremediation,* the name of the new discipline that studies how biological processes can be used to destroy or neutralize pollutants.

More germane to the subject at hand, microbiologists have discovered that *halogenated compounds*—compounds that contain chlorine, bromine, fluorine, or iodine—are a favorite food and energy source for some soil bacteria. Extensive laboratory studies have demonstrated the existence of families of bacteria that produce enzymes to break up and eat these halogenated compounds. The number of such man-made halogenated organic chemicals is extensive, including most pesticides, herbicides, and insecticides; many other industrial solvents and chemicals; and all chemicals banned under the Montreal Protocol, including chlorofluorocarbons (CFCs).

Bacterial colonies have been cultivated and grown in the laboratory and successfully used in field experiments in hazardous waste sites where the bacteria have completely destroyed deposits of carbon tetrachloride (a suspected carcinogen) and other halogenated compounds. Carbon tetrachloride, it should be noted, is one of the chemicals banned under the Montreal Protocol as an alleged ozone depleter.

Microbiologists who have studied these dehalogenating bacteria for years say there is no reason that these bacteria cannot also destroy CFCs. One team of microbiologists has gone a step further, and tested soils to see if bacteria are destroying CFCs; their laboratory experiments have fully proven that it happens. (See Figure 4.2.)

Soil bacteria are not the only sink for CFCs in the atmosphere, however; there are others.

Perhaps the best starting point is the study that led microbiologists to look at how bacteria and bacterial enzymes destroy CFCs. Six years ago, in 1986, Aslam Khalil and Reinhold Rasmussen, two of the world's leading atmospheric chemists, were taking measurements of methane and carbon dioxide emissions from termite mounds in Australia. The scientists were also measuring CFCs as tracers in order to calibrate their instruments, because CFCs were supposed to be indestructible. To their great surprise, however, they discovered that something—or some process—in the soil was destroying the CFCs.

Khalil and Rasmussen, who work out of the Institute of Atmospheric Sciences at the Oregon Graduate Center, reported their

Figure 4.2
SOILS DESTROY SIGNIFICANT AMOUNTS OF CFCs

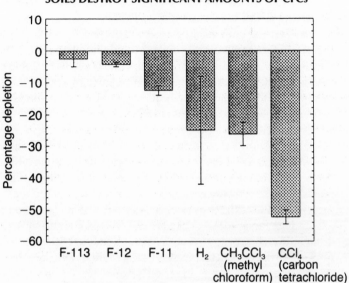

Trace gases (25 centimeters below the surface)

Significant depletion of CFCs—especially methyl chloroform and carbon tetrachloride—occurs a short distance below the soil surface, as this graph shows. For instance, the concentration of carbon tetrachloride just 25 centimeters below the surface of the soil is only 50 percent that of the ambient air concentration. As yet, the processes destroying the CFCs are unknown. It is possible that certain types of soil have microorganisms that scavenge chlorine from CFCs to use metabolically. Several scientists interested in pursuing this line of research have had their requests for funding rejected.

Source: Adapted from M.A.K. Khalil and R.A. Rasmussen, "The Potential of Soils as a Sink of Chlorofluorocarbons and Other Man-Made Chlorocarbons," *Geophysical Research Letters*, Vol. 16, (July 1989).

results in the July 1989 issue of *Geophysical Research Letters* in a paper titled "The Potential of Soils as a Sink of Chlorofluorocarbons and Other Man-Made Chlorocarbons." Their measurements, they said, showed amazingly rapid removal of chlorocarbons by the soils and other constituents of the termite mounds. The soils depleted methylchloroform (CH_3CCl_3) and carbon tetrachloride (CCl_4) by about 25 percent and more than 50 percent, respectively. According to their paper, "Such large changes cannot be explained by the slowness of the processes that transport these chemicals into the soils of the termite mounds or soils in general. . . . [It] can only be a result of either adsorption in the soil or removal by heterogeneous or biological processes" (p. 680).

Khalil and Rasmussen also observed rapid removal of CFCs, but not in as great magnitudes as the chlorocarbons. Nevertheless, soils could still potentially deplete 15 percent of certain CFCs annually. (These are measured local depletions; to calculate destruction rates is very complicated.) Their paper concludes:

> We have shown that soils remove man-made chlorocarbons. Whether the soils act as passive reservoirs or chemically convert the chlorocarbons to other compounds cannot be determined from these experiments. It should be noted that the soils of the earth have been exposed to the chlorocarbons we studied for at least a decade, and they still continue to remove the chlorocarbons at measurable rates. The rates of removal are greatest for carbon tetrachloride followed by methylchloroform and F–11 [p. 682].

The breakup of CFCs and other chlorocarbons banned by the Montreal Protocol occurs in rice paddies as well. On another field trip, Khalil and Rasmussen traveled to China, where, in collaboration with the Chinese Academy of Sciences and two Chinese scientists, they measured the emission rates of methane and other gases from rice fields and biogas generators. In all, they conducted about 30 experiments and obtained some 900 individual measurements of trace gases emitted from rice fields and biogas generators around Chengdu in Sichuan province of China.

As in the Australian expedition, Khalil and Rasmussen took careful measurements of CFC absorption by the rice paddies. In their paper on the expedition, published in *Chemosphere* in 1990, they conclude: "There is evidence . . . that the rice fields take up some man-made chlorocarbons, particularly CFC–11, CFC–12 [carbon tetrachloride], and [methylchloroform]. . . . [T]hese gases are implicated in the depletion of the ozone layer" (p. 207).

This last comment is significant, coming from two of the world's leading atmospheric chemists. Aslam Khalil carried out some of the most thorough measurements of CFC concentrations in the atmosphere in the early 1980s, and his work is cited as the standard source on concentrations of CFCs in the atmosphere, while Rasmussen has spent the past 10 years setting up and calibrating most of the CFC monitors around the world. Both scientists realized that there are major problems with the ozone depletion theory, and they are willing to examine these problems.

The paper reviewing the results of their work in China (Khalil et al., 1990b) concludes:

> In our recent studies we have found that soils tend to take up man-made chlorocarbons particularly CCl_3F (F–11) and CCl_4 (carbon tetrachloride) (Khalil and Rasmussen, 1989). When we looked at the fluxes of chlorofluorocarbons in the rice field experiments, we found once again that the fluoro-carbons were being taken up by the paddies. . . . [T]he mean rates of removal of F–11, F–12, CCl_4, and CH_3CCl_3 are of the same order of magnitude but several times higher than found in studies on Australian soils. . . [p. 225].

In another paper on the influence of termites on atmospheric gases [1990a], Khalil and Rasmussen point out that termite mounds are major sources of atmospheric gases like carbon dioxide and methane, as well as chloroform ($CHCl_3$). Other scientists have commented that it is very possible that some of the CFCs that are disappearing in the termite mounds are actually being broken up by bacteria in the gut of the termites. The chlorine atom is then used by the bacteria as either food or energy and is emitted into the air as chloroform.

Why CFCs Attract Bacteria

Why would bacteria that live in oxygen-poor (anaerobic) environments want to break up CFCs and other man-made halogenated compounds? There are two basic reasons.

First, bacteria, like all other living organisms, including man, need to "respire" and to obtain energy. In the presence of oxygen, bacteria pass on an electron obtained from a food source to an oxygen molecule, creating a water molecule. This is the basic act of respiration. In the absence of oxygen, bacteria have a problem because they have to find a receptor for that electron. In swamps and oxygen-poor soils, the bacteria that predominate are those that are specially capable of using other compounds to which they can transfer electrons and which serve as a source of carbon. The favorite compounds that replace oxygen are reduced forms of nitrogen, sulfur, and the halogens.

The second reason that bacteria are attracted to the halogenated compounds is that the carbon atoms in halogenated molecules represent a source of food, and bacteria are constantly in search of food.

How common is the bacterial breakdown of halogenated compounds? Joseph Suflita, professor of microbiology at the University of Oklahoma, commented on Rasmussen and Khalil's discovery: "[T]he removal of halogens from organic molecules seems to be favored under anaerobic conditions, which is the kind of conditions you have in rice paddies and termite mounds. So without a doubt, their discovery makes a certain amount of sense."

Suflita, one of the leaders in environmental microbiology, described in a March 1991 interview with one of the authors how the process works:

> Mechanistically, bacteria use the halogenated organic compounds as a terminal electron receptor. They respire with it, just the same way you and I respire with oxygen as our terminal electron receptor. The process is called a reductive dehalogenation, which literally means that a halogen is removed and substituted by a proton or a hydrogen atom. So one of the products would be a halogenated intermediate with one less halogen than the starting material. Then that could undergo subsequent dehalogenation reactions also.
>
> So as a replacement of that chlorine, or maybe in the case of CFCs, the fluorine, by a proton, the chlorine comes out as inorganic chlorine, as hydrochloric acid. Therefore, it is converted from an organic molecule to an inorganic molecule.

The extensive laboratory work on carbon tetrachloride documents this process. Scientists have observed that in the absence of oxygen, soil bacteria will break up the carbon tetrachloride (CCl_4) in a stepwise fashion. First, one chlorine atom will be broken off, enabling the bacteria to deposit one electron on the chlorine atom, creating HCl, and one on the CCl_4 molecule, creating $CHCl_3$. In turn, each one of the remaining chlorine atoms is broken off, so that by the time the bacteria are finished, what is left is a molecule of methane CH_4 and four HCl molecules.

Although this mechanism has been well established by years of laboratory work and field observations, there is a disagreement among scientists as to what ultimately happens to the now completely dehalogenated molecule. While Suflita maintains that the bacteria do not use the remaining carbon atom as a food source, Fred Brockman from Battelle Pacific Northwest Laboratories maintains that they do (interview, 1991).

Brockman and Suflita agree up to the point that a methane molecule, CH_4, is created. At that point, the Battelle scientists have observed that the bacteria then oxidize the methane molecule, turning it into a carbon dioxide molecule, CO_2. This process, according to Brockman, is extremely important for bacteria where food (carbon atoms) is scarce. The bacteria will break up the halogenated compounds to respire and ultimately to consume the carbon atom.

While most microbiologists have been studying the bacterial destruction of the more common halogenated compounds found in toxic waste sites, one team of scientists has examined the destruction of CFCs by these bacteria. Provoked by the papers published by Rasmussen and Khalil, Derek Lovely and Joan Woodward, working at the U.S. Geological Survey in Reston, Virginia, spent a year looking at what happens to CFCs in anaerobic soils. They announced their results in a paper at the December 1990 conference of the American Geophysical Union in San Francisco, "Consumption of Freons CFC–11 and CFC–12 in Methane-Producing Aquatic Sediments."

Lovely and Woodward took swamp sediments from the Potomac River to the laboratory and introduced Freons into the sealed vials. They discovered that within a short period of time all the CFC–11 and CFC–12 had disappeared. When they heated the sediments to kill all biological organisms, they found that although CFC–12 would no longer be absorbed by the sterilized soil, CFC–11 would still be consumed, albeit at much slower rates than if bacteria were present. Their conclusion is that both biological and abiological processes in the soils consume CFC–11, whereas CFC–12 is consumed only by biological mechanisms.

The Lovely and Woodward findings are entirely coherent with the behavior of bacteria that destroy halogenated compounds, as observed and measured by other scientists. As of this writing, the researchers had not calculated the global rate of CFC destruction by bacteria, but from all the evidence available it is quite significant.

In addition to bacteria, other chemicals and minerals also destroy CFCs through catalytic reactions, according to Ron Crawford, codirector of the Center of Hazardous Waste Remediation Research at the University of Idaho. Crawford's observations, along with those of Lovely and Woodward, also cohere with the findings of Dean Hegg et al. (reported below) that CFC–12 is found in high concentrations in the smoke plumes of forest fires but CFC–11 is not. The fires studied by Hegg and his associates occurred in soils that were very rich in oxygen. Under these circumstances, the predominant type of bacteria would be those that most effectively utilize oxygen in their metabolism. With such an abundance of oxygen, there would be no need for the extra effort required of the bacteria to break up the CFCs. Also, those families of bacteria that specialize in breaking up halogenated compounds would not be abundant. Under these conditions, the abiological processes can destroy CFC–11, while the biological processes would be much slower than in anaerobic environments.

Fred Brockman of Battelle Laboratories has pointed out that one of the most interesting capabilities of microbes is their ability to evolve to meet new environmental conditions and new food sources: They are constantly scanning the environment for new food sources and will actually migrate when they detect new sources elsewhere. Furthermore, the microbes have specialized enzymes that can be turned on as necessary to break up different kinds of compounds, and certain microbes are better at utilizing certain new sources of food.

Along these lines, Crawford notes that at the bottom of the Hudson River there are colonies of microorganisms that consume PCBs. Tests at the bottom of lakes where there never was any PCB contamination show that such organisms are not found there. This indicates to the microbiologists that either those organisms were present in only very small quantities and multiplied with the arrival of the new food source, or that they evolved to utilize the new food source more efficiently.

This last point could drastically change how we look at waste and pollutants in the future: Microbiologists are now capable of nurturing microbes that can clean toxic waste sites at a fraction of the present cost of cleanup. And in a few years, soil bacteria across the world may evolve to the extent that more CFCs will be destroyed than are produced. At that point, the concentrations of this new food source—CFCs—in the atmosphere will not only stabilize, but start to decrease!

Where There Is Smoke, There Are CFCs

Destruction by bacteria and bacterial enzymes is not the only thing happening to CFCs in the ground. Another fascinating discovery was made by Dean Hegg and his collaborators at the Department of Atmospheric Sciences, University of Washington. Using the university's C–131A research aircraft, the scientists conducted studies of gases emitted by forest fires at seven different locations across the United States. They measured the standard gases expected to come from forest fires and, at the same time, they also took measurements of CFCs. To their great surprise, the scientists discovered the forest fires were emitting huge volumes of CFCs.

This is rather extraordinary, since trees don't produce CFCs, only man does. Furthermore, the volume of CFC–12 being emitted by the forest fires would account for fully 50 percent of all the Freons in the atmosphere. The only explanation, the scientists conclude, is that CFC–12, because it is so heavy, has been absorbed on the soil surface and on plant debris and the fires are lifting it back to the atmosphere in the smoke column. As the scientists report in the April 1990 *Journal of Geophysical Research,* "the high emissions of NO_x and CFC–12 are due in whole or part due to the resuspension of previously deposited pollutants" (p. 5669).

There is another explanation, however. One of the reasons that CFCs are not readily absorbed by the oceans and other bodies of water is that they have low solubility. However, CFCs are 300 to 400 times more soluble in lipoproteins than in water. Therefore, it is possible that lipoproteins in plants are sequestering CFC molecules and that there is a large reservoir of CFCs in biomass. The Freon–12 observed in the smoke plumes from the forest fires could have been coming from the plant tissues, not from the soils.

Hegg et al. (1990) write:

> The only sink for F12 that is generally considered in atmospheric budget calculations is loss by photodisassociation in the stratosphere [National Research Council 1983]. However, the global flux of F12 from biomass burning . . . (200,000 tons per year) is >50% of the estimated yearly global emissions

of F12 (>400,000 tons per year). [Therefore], since F12 cannot be produced by fires, it must have been previously deposited onto the fuel bed and revolatized in the fires. Hence, our results suggest that deposition of F12 may be important on the global scale, contrary to current understanding of this issue. . . [p. 5673].

Although high concentrations of Freon were found in all forest fires studied, the highest concentration occurred at the site called Lodi I near Los Angeles. The fire that was studied happened six days after it had rained, which could mean that particulates in falling raindrops had captured a large number of Freon molecules and deposited them. Estimating that their Freon–12 readings in the smoke column were perhaps abnormally high because of these conditions, the scientists then eliminated all the Lodi I data and used only the measurements from the other six forest fires in their calculations. Nevertheless, even without including the high emissions from Lodi I they found that the emissions from other fires still yield "a global flux of >60,000 tons of F12 from biomass burning worldwide. . . . This is still quite significant (15% of the worldwide flux of F12)," they say.

The paper concludes that "the deposition of F12 onto the Earth's surface could be the tropospheric sink postulated as a possible explanation for discrepancies in F12 atmospheric lifetime estimates" (p. 5673).

The results that Hegg et. al., obtained were for soils exposed to fires, which account for only a small proportion of the Earth's surface. If such a huge amount of CFCs (15 to 50 percent of the Freon–12 in the atmosphere) is being emitted by forest fires in such a small portion of the Earth's surface, what about the amount of CFCs in the rest of the soils across the world? In fact, no one has ever examined the soils to see if CFCs were drifting downwards and just staying there. Instead, hundreds of millions of dollars have been spent sending balloons and high-flying aircraft to the stratosphere.

Anyone who has ever worked with CFCs knows that these chemicals do not rise easily in air. A keen description of this phenomenon is given by Bob Holzknecht, president of the Automotive Air Group, a chain of 400 independent shop owners who repair auto air-conditioners in the United States. Holzknecht is a mechanical engineer who has spent more than 20 years experimenting, lecturing, and writing about the technical problems (and their solutions) facing professional air-conditioner repairmen. In the June 1990 issue of the the group's technical newsletter, *The Accumulator,* Holzknecht scathingly attacks the ozone depletion theory:

> Leaking freon 12 does not waft out nor disperse throughout the atmosphere. It stratifies sharply and immediately. The molecules fall straight down toward the floor. In our enclosed shop, there is no buildup except the bottom of a depression in the floor. Swirling low level air currents push them along as they seek lower and lower resting places. . . .
>
> I can physically demonstrate in my own auto-repair shop that: freon 12 does not rise, freon 12 does not waft thru the air; that freon 12 does not mingle with air; freon 12 stratifies tightly and displaces air [p. 1].

Holzknecht's crucial experiment for the skeptics is simple but effective: "Fill an ordinary water glass with freon 12. Place it uncovered on an open shelf in a relatively still corner. Next day, next week, next month, next year, indefinitely; most of that freon 12 will still be inside that open water glass."

Confirming Holzknecht's description, a researcher for one of the largest U.S. medical supply companies pointed out to one of the authors that the company has had Freon in an open container for more than 15 years now, and all of it is still sitting there.

What About the Oceans?

If CFCs are so heavy that they tend to just sit in the ground, what about the oceans, which cover three-quarters of the surface of the Earth? In fact, despite their low solubility, CFCs are being absorbed into the oceans, and the National Oceanic and Atmospheric Administration (NOAA) is now engaged in a worldwide study to determine ocean concentrations of CFCs. The purpose of the study is to use the CFCs as tracers for deep water currents. It takes decades for water from the surface of the ocean to be transported to the bottom, and vice versa. CFCs are excellent tracers in this case because they have been in use for less than 50 years, and the ratio of different CFCs in use at different times can be used to date when the CFCs were absorbed by the surface waters.

Of great interest here, however, is the fact not only that CFCs are being absorbed by the oceans, but also that they have been detected at depths of 6 kilometers under the surface. If CFC molecules are so adept in rising to the stratosphere, what are they doing at the bottom of the ocean?

Figure 4.3 shows concentrations of CFC–12 dissolved along a section of the North Atlantic, as published in the November 1989 *Oceanography* in an article by NOAA oceanographer John Bullister (p. 15). Although the concentrations of CFC–12 measured in the oceans are low, this represents a vast reservoir for CFCs, since the oceans hold an enormous volume of water.

One further question has to be asked: Are CFCs perhaps being destroyed at the surface of the ocean? A very interesting hypothesis on this by Judith Sims et al. appears in a paper published in the Winter 1990/91 issue of the journal *Remediation:**

> Large-scale production of synthetic halogenated organic compounds, which are often resistant to both biotic and abiotic degradation, has occurred only in the last few decades. . . . However, naturally occurring halogenated organic compounds have existed in marine systems for perhaps millions of years. These compounds, including aliphatic and aromatic compounds containing chlorine, bromine, or iodine, are produced by macroalgae and invertebrates. The presence of these natural compounds, at potentially high concentrations, may have resulted in populations of bacteria that are effective dehalogenators. . . [p. 75].

In other words, the greatest concentration of CFC-destroying organisms may be present at the surface of the world's oceans. It should be further noted that many of the halogenated compounds produced by living organisms in the oceans are identical to most man-made chlorocarbons, while some are very similar to chlorofluorocarbons (CFCs). As a matter of fact, several scientists have pointed out that there are organisms in the oceans that manufacture fluorinated compounds. This raises two further issues. First, as noted by Sims et al., entire bacterial colonies have evolved the

---

* For those interested in more details on this subject, this is an excellent review paper. A table in this article presents the classes of halogenated compounds that have been observed to be degraded by reductive dehalogenating processes in the soils. The paper also defines oxidation and reduction reactions in detailed but understandable terms for the layman.

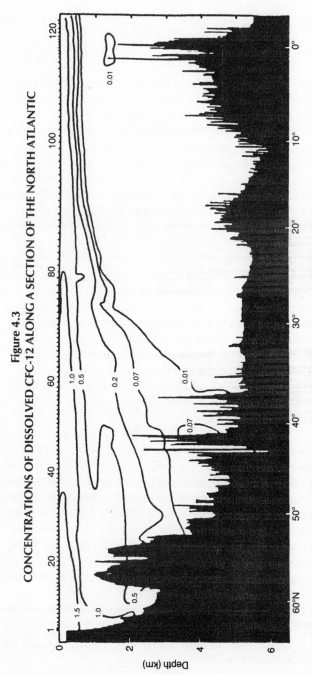

Figure 4.3

CONCENTRATIONS OF DISSOLVED CFC-12 ALONG A SECTION OF THE NORTH ATLANTIC

More than 2,000 samples along this section of the ocean floor were analyzed for CFC content. If CFCs are destroying the ozone layer in the stratosphere, as the ozone depletion theorists claim, then what are these CFCs doing at the bottom of the ocean?

Source: Adapted from John L. Bullister, "Chlorofluorocarbons as Time-Dependent Tracers in the Ocean," *Oceanography*, November 1989.

capability of utilizing halogenated compounds as sources of energy and food. There is no reason that these colonies would not be able to use CFCs for that purpose. Second, because these natural fluorocarbons are nearly identical to CFCs, is it possible that they have been confused with CFCs in the atmospheric measurements? Both issues should be investigated.

References

Fred Brockman et al., 1989. "Isolation and Characterization of Quinoline-Degrading Bacteria from Subsurface Sediments," *Applied & Environmental Microbiology,* Vol. 55, No. 4, pp. 1029–1032.

John L. Bullister, 1989. "Chlorofluorocarbons as time-dependent tracers in the ocean," *Oceanography* (November) pp. 12–17.

P. Fabian, R. Borchers, S.A. Penkett, et al., 1981 "Halocarbons in the Stratosphere," *Nature* (Dec. 24), pp. 733–735.

P. Fabian, R. Borchers, G. Gömer, et al., 1984. "The Vertical Distribution of Halocarbons in the Stratosphere." In *Atmospheric Ozone,* Proceedings of the Quadrennial Ozone Symposium, Sept. 1984. Eds. C. Zerefos and A. Ghazi. Dordrecht, Netherlands: Reidel.

Dean A. Hegg, Lawrence F. Radke, Peter V. Hobbs, et al., 1990. "Emissions of Some Trace Gases from Biomass Fires," *Journal of Geophysical Research,* Vol. 95, No. D5 (April 20), pp. 5669–5675.

Robert Holzknecht, 1990. "That Vicious Ozone Depletion Propaganda," *The Accumulator* (June), pp. 1–4.

Aslam Khalil and R.A. Rasmussen, 1989. "The Potential of Soils as a Sink of Chlorofluorocarbons and Other Man-Made Chlorocarbons," *Geophysical Research Letters,* Vol. 16, No. 7 (July), pp. 679–682.

Aslam Khalil, R.A. Rasmussen, J.R.J. French, et al., 1990. "The Influence of Termites on Atmospheric Trace Gases: CH_4, CO_2, $CHCl_3$, N_2O, CO, H_2, and Light Hydrocarbons," *Journal of Geophysical Research,* Vol. 95, No. D4 (March 20) pp. 3619–3634.

Aslam Khalil, R.A. Rasmussen, M.Y. Wang, et al. "Emissions of Trace Gases from Chinese Rice Fields and Biogas Generators: CH_4, N_2O, CO, CO_2, Chlorocarbons, and Hydrocarbons," *Chemosphere,* Vol. 20, No. 1–2, pp. 207–226.

Derek Lovely and Joan Woodward, 1990. "Consumption of Freons F–11 and F–12 in Methane-Producing Aquatic Sediments." Paper presented at the fall meeting of the American Geophysical Union, San Francisco, Calif., (Dec. 3–7).

F. Sherwood Rowland and Mario Molina, 1975. "Chlorofluoromethanes in the Environment," *Reviews of Geophysics and Space Physics,* Vol. 13, No. 1, pp. 1–35.

F. Sherwood Rowland, 1989. "Chlorofluorocarbons and the Depletion of Stratospheric Ozone," *American Scientist,* Vol. 77. No. 1, p. 36.

Judith L. Sims, Joseph M. Suflita, and Hugh H. Russell, 1991. "Reductive Dehalogenation: A Subsurface Bioremediation Process," *Remediation*, Vol. 1, No. 1, pp. 75–93.

Peter Warneck, 1988. *Chemistry of the Natural Atmosphere*, Vol. 41: International Geophysics Series. New York: Academic Press.

S.C. Wofsy and M. B. McElroy, 1973. "On Vertical Mixing in the Upper Stratosphere and Mesosphere," *Journal of Geophysical Research,* Vol. 78, p. 2619.

5

The Antarctic Ozone Hole

After aerosol cans were banned in the United States in 1978, the CFC issue lay dormant for several years. It was revived dramatically in 1985, when Joseph Farman of the British Antarctic Survey announced the discovery of a gaping "hole" in the ozone layer over Antarctica. Suddenly, the press had banner headlines claiming that the ozone hole, "larger in area than the United States," was expanding to menace the entire world. The ozone depletion theorists sprang into action. It did not matter that the Rowland/Molina ozone-depletion theory neither predicted the existence of the "hole" nor could possibly explain it. The Antarctic ozone hole became the proof that the depletion theorists had been looking for.

Concealed from the public in all the hoopla was the fact that the so-called ozone hole was not new; its existence had been known for more than 30 years. Scientists had discovered what they called at the time the southern anomaly in the ozone layer in 1956–1957, during the International Geophysical Year (IGY), when ozone spectrophotometers were placed in Antarctica for the first time. The discoverer was ozone research pioneer Gordon Dobson, who coordinated the British team at Halley Bay, Antarctica.

In an article titled "Forty Years' Research on Atmospheric Ozone at Oxford: A History," published in *Applied Optics* magazine in March 1968, Dobson describes how the anomaly came to light:

One of the most interesting results on atmospheric ozone which came out of the IGY was the discovery of the peculiar annual variation of ozone at Halley Bay. This particular ozone instrument [was sent] to Shotover [laboratory] to be checked up immediately before leaving England. Moreover, Evans, who took the original observations at Halley Bay, had also been to Shotover to become familiar with the working of the instrument and its maintenance. The annual variation of ozone at Spitzbergen [near the North Pole] was fairly well known at that time, so, assuming a six months difference, we knew what to expect. However, when the monthly telegrams from Halley Bay began to arrive and were plotted alongside the Spitzbergen curve, the values for September and October 1956 were about 150 [dobson] units lower than was expected. We naturally thought that Evans had made some large mistake or that, in spite of checking just before leaving England, the instrument had developed some fault. In November the ozone values suddenly jumped up to those expected from the Spitzbergen results. It was not until a year later, when the same type of annual variation was repeated, that we realized that the early results were indeed correct and that Halley Bay showed most interesting difference from other parts of the world. It was clear that the winter vortex over the South Pole was maintained late into the spring and that this kept the ozone values low. *When it suddenly broke up in November both the ozone values and the stratosphere temperatures suddenly rose* [emphasis added] [p. 403].

Figure 5.1 illustrates the measurements Dobson describes. His description of the "peculiar variation" suggests the ozone hole. However, the normal winter values of total ozone over Antarctica are around 250 dobson units. That is, Dobson did not detect the steep September–October dip we now identify as the ozone hole. But, as noted below, the French, at their Antarctic station at Dumont d'Urville in 1958, did detect a steep September–October dip, down to values as low as 110 dobson units.

The polar vortex—an enormous vortical structure—does not stay in one spot. It may move several times a month and moves

121

Figure 5.1
**DOBSON'S OZONE OBSERVATIONS AT HALLEY BAY,
ANTARCTICA, 1956-1959**

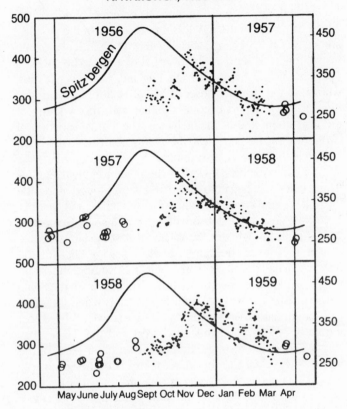

o = Moon observations

*Dobson's original caption reads: "The full curve is for Spitzbergen
[near the North Pole], shifted by six months. Note the lower values
of ozone in the southern spring and the sudden increase in November at the time of the final atmospheric warming." The Antarctic
data are shown as dots, each representing a reading of ozone layer
thickness. The amount of ozone takes a huge leap at the end of
October, as can be seen, when the polar vortex breaks up.*

Source: *Applied Optics*, March 1968. Vol. 7, No. 3.

many times in the course of a year. An atmospheric station near the edge of Antarctica may be inside the vortex one year and outside the next year, and ozone concentrations will vary dramatically. Inside the vortex, the values measured may be as low as 110 dobson units, while just a few miles away, outside the vortex, the values may be as high as 450 dobson units. These facts answer the main argument of the ozone depletion theorists that man-made CFCs are responsible for the increase in the southern anomaly, the so-called ozone hole.

Two French scientists, P. Rigaud and B. Leroy, recently republished the 1958 data from the French Antarctic station, Dumont d'Urville, located on the opposite side of the South Pole, a few hundred miles away from Halley Bay. These measurements show that the ozone hole was deeper in 1958 than at any time in the 1980s, but that it disappeared immediately after the breakdown of the 1958 polar vortex.

The Rigaud-Leroy paper, which appeared in the November 1990 issue of *Annales Geophysicae,* reports on the scientists' search through old ozone records from the French Antarctic Observatory going back to 1958. Rigaud and Leroy discovered that ozone levels took a precipitous decline at the beginning of 1958's austral spring, August and September, and reached values of as low as 110 dobson units—the lowest values ever recorded in Antarctica! These data were recorded and published in the scientific literature in the 1960s, but had not been examined recently.

The team concluded that, "reexamination of the [Dumont d'Urville] data shows that a strong minimum of the total ozone content has been observed that year in the austral spring time. This suggests a natural phenomenon to explain the Antarctic 'ozone hole' " (p. 791). According to the scientists, the "ozone hole" appears in September and the beginning of October 1958, but then there is "a spectacular recovery of the ozone concentration between October 8 and 21. The polar vortex breakdown in 1958 occurred between October 5 and 20," they report (p. 793).

What could explain this dramatic drop to 110 dobson units recorded at the Dumont d'Urville station, while at Halley Bay the readings were around 250 dobson units? Rigaud and Leroy write:

> The center of the polar vortex [was] near Dumont d'Urville at the end of winter [1958] and far from Halley Bay. The situation is the opposite of the one observed in the recent

years. Since the "ozone hole" is observed in the polar vortex, this could explain why this phenomenon was undetected in 1958 at Halley Bay [p. 793].

In other words, the polar vortex was in a completely different location in 1958 from where it is today. The French data show that although the values of ozone were not that low at Halley Bay, in another part of the vortex, even farther from the pole, the values of ozone dropped precipitously to values as low as—and in many cases significantly lower than—those being recorded today. This is, we should remind the reader, 34 years ago, when CFCs were barely in use.

The French scientists also put forth an interesting hypothesis: that the cause of the so-called ozone hole has to do with the optical properties of light traveling through the atmosphere:

> At Dumont d'Urville atmospheric illumination is crepuscular for a long time. The sun culminates at $+5°$ of elevation on August 1 and at $+15°$ on September 1st. In this case the destruction cycles of the ozone are very important because of the successive arrival of the different wavelengths of the solar radiation.
>
> As shown by Hoffman and Rosen (1985), the major volcanic eruptions affect the Antarctic stratospheric aerosol layer. It is known that such large eruptions took place in recent years at St. Helens in 1980 and at El Chichón in 1982, but also at Bezymianny [Siberia] in 1956, 2 years before the measurements made at Dumont d'Urville. The perturbations of this aerosol layer could therefore partly explain the "ozone hole" owing for example to an unknown heterogeneous chemistry or to a change in the illumination of the twilight and therefore to a change in the photodissociation rates of the species.

In other words, depletion of ozone could be caused by changes in stratospheric chemistry brought on by large volcanic explosions, either through chemical changes in the vortex itself or changes in the wavelengths of light as they travel through the Earth's atmosphere at an oblique angle, before they strike Antarctica.

Rigaud and Leroy conclude:

Reexamination of the ozone spectrographic data obtained at Dumont d'Urville in 1958 shows that the "ozone hole" was already present that year in September.

Although chlorofluorocarbon production was already increasing in 1958, its abundance was far from the concentration today. Therefore, the existence of an Antarctic ozone depletion above Dumont d'Urville in September 1958, suggests that natural phenomena such as volcanic eruptions also contribute to ozone destruction [pp. 793–794].

It is true that the levels of ozone over Antarctica at the beginning of the 1958 austral spring appear to be lower today than those measured during the IGY. However, that does not mean that man-made CFCs have depleted the ozone layer. For decades, it has been known that the ozone layer follows several long-term cycles, including the solar cycle, and the thickness of the ozone layer can vary significantly depending on these cycles. In Antarctica, in addition to the long-term cycles of ozone determined by the solar phenomena, atmospheric dynamics plays a major role in the formation and length of the so-called ozone hole, including the polar vortex and the complex chemistry of the polar atmosphere.

At the end of the polar night and beginning of the austral spring, an intense, hurricane-like vortex surrounds Antarctica. With the jet stream traveling at speeds greater than 200 miles per hour (mph), the vortex effectively seals the air inside Antarctica from the air outside. Enormous quantities of ozone, produced in the tropics and drifting toward the poles, are stopped by the polar vortex. As Figure 5.2 shows, there are huge concentrations of ozone on the outside of the polar vortex, while at the same time ozone levels are dropping inside the vortex. The polar vortex breaks up after a few weeks, allowing the ozone-rich air from the tropics to replenish the ozone layer over Antarctica. This raises the ozone levels to those measured by Dobson in 1956–1958 in late spring and summer.

In other words, the ozone hole has been an observed annual phenomenon since about 1979, lasting between four and six weeks of the year-at the period of transition from the darkness of winter until the spring breakup of the polar vortex. The rest of the year, Antarctic ozone values are quite similar to those measured during the IGY 36 years ago. However, the observations prior to 1979, showing no hole, and the French observations at Dumont d'Urville

Figure 5.2
DEPTH OF ANTARCTIC OZONE HOLE IN 1987

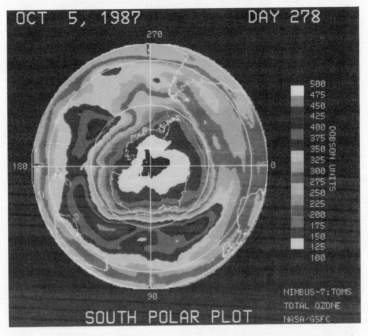

This photograph, generated by computer from data gathered by NASA's Nimbus 7 satellite, shows the concentrations of ozone during the occurrence of the Antarctic ozone hole in 1987. The bar at right shows the relative concentrations of ozone. There are low values of ozone in the center of the plot, but this Antarctic ozone hole is surrounded and sealed off by a surrounding band of air that has unusually high concentrations of ozone. This enormous "mountain" of ozone surrounding Antarctica ranges in thickness from 300 to 475 dobson units, significantly higher concentrations of ozone than at the tropics, as the same plot reveals. This phenomenon develops because very little ozone is produced at the poles; most of it is produced in the tropic latitudes and then is moved by planetary waves and other dynamic phenomena in the atmosphere to the poles. In early autumn, the poles are sealed off from the rest of the atmosphere by the polar vortex, preventing ozone from entering the Antarctic atmosphere. By November, however, the polar vortex breaks up, and the "hole" is filled.

Source: NASA Goddard Space Flight Center

in 1958 suggest that the ozone hole is ephemeral as suggested by Dr. S. Fred Singer. That is, it comes and goes depending perhaps on variations in stratospheric temperatures, water vapor, or chlorine from volcanic eruptions.

From a rigorous scientific standpoint, there is no need to go any further to disprove the theory that CFCs are responsible for causing this so-called Antarctic ozone hole, because the hole is proven to have been there decades before man-made CFCs could possibly have had any impact. The CFC ozone depletion theory is a fraud. Unfortunately, however, ozone-depletion theories share the un-canny ability of the cinema character Dracula; both can reappear in a new incarnation after having been thoroughly crushed. There-fore, let us continue to rebut every aspect of this hoax to establish without a doubt that CFCs are at most a negligible addition to natural processes that have been able to produce an ozone hole without help from man at least as early as 1958—and likely for the past 50 million years since the Antarctic continent has been located at the South Pole.

The Japanese Evidence

Some of the best research on the Antarctic atmosphere has been carried out by Japan's National Institute of Polar Research. In fact, one of Japan's leading scientists, Shigeru Chubachi, is the actual discoverer of the recent increase in the southern anomaly, the so-called ozone hole. Chubachi noted an increase in the southern anomaly three years before Joseph Farman of the British Antarctic Survey published his paper in 1985 (Farman et al.).

The story of the rediscovery of the Antarctic ozone hole began in 1982, when Japanese scientists operating the Antarctic station at Syowa (69°S, 40°E) announced that they had observed a drop in the average values of ozone in Antarctica every year for several years. This discovery was reported in a scientific paper in 1983. In September 1984, Chubachi gave a presentation at a world confer-ence on ozone in Athens, Greece, attended by a large number of atmospheric scientists. His paper (Chubachi 1985), written in English, plainly stated that the Japanese had detected a drop in ozone values at Syowa in 1982.

Eight months later, British scientist Joseph Farman announced his alleged discovery of the ozone hole in a May 1985 article in *Nature* magazine.*

Two scientists from Tokyo's National Institute of Polar Research have argued that the ozone hole is created by atmospheric dynamics. Hiroshi Kanzawa and Sadao Kawaguchi write, in an elegant paper published in the January 1990 issue of *Geophysical Research Letters,* that "the data show that dynamics plays an essential role in many aspects of the Antarctic ozone hole phenomenon" (p. 77).

At Syowa Station the Japanese observed that "a large sudden warming occurred in 1988 from the end of August to the beginning of September" in the polar stratosphere. Then, "during the same period as the sudden warming ... there occurred a sudden increase of total ozone over Syowa Station...." This was followed by a sudden cooling of the stratosphere, at which point "there occurred a sudden decrease of total ozone during the same period of the sudden cooling...." A further important point, the Japanese scientists say, is that "as the warming event progressed, the ozone-rich region equatorward of Syowa Station with total ozone amount larger than 400 dobson units (DU) moved poleward..." (p. 77).

In other words, despite the fact that the Antarctic atmosphere was chemically "primed" with loads of chlorine—which according to the ozone-depletion theory should have gobbled up every ozone molecule for miles around—the Japanese recorded the highest ozone readings for that period in almost two decades. This was because the warming of the stratosphere by 10° was followed by the movement of ozone-rich air from outside the polar vortex, a process shown clearly in Figure 5.3. The Antarctic ozone hole had started to develop by Aug. 18; but then warm air moved into the region of the vortex; by Aug. 31 the ozone readings jumped from 236 units to 463 units. By Sept. 9, when the stratosphere had cooled again, the ozone layer was depleted to levels similar to those recorded on Aug. 18.

The Japanese scientists, who have examined their data much

---

* Despite the fact that scientists who follow the field have pointed out that Chubachi should receive the credit, Farman continues to claim he is the discoverer of the ozone hole. One scientist interviewed by the authors pointed out that for reasons of racial prejudice, Japanese and other scientists from non-Western nations, like India, receive little credit for the scientific work they do. He alleged that papers written in Japanese or other languages are easily plagiarized, since so few people in the West read these languages.

Figure 5.3
OZONE ABUNDANCE IS DIRECTLY RELATED TO TEMPERATURE

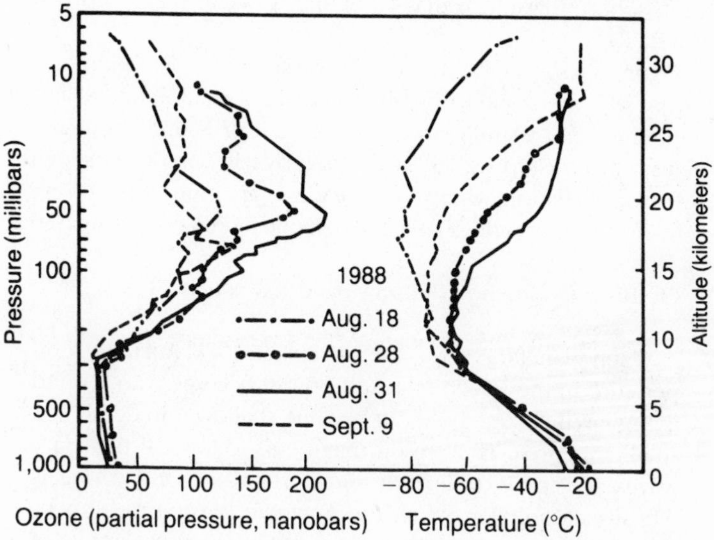

The so-called ozone hole appears when the stratosphere is cold and disappears when it warms up. The abundance of ozone is directly related to the temperature of the stratosphere, as can be seen in these vertical profiles of ozone (in partial pressure) and temperature, measured at Antarctica's Syowa Station in 1988. On the days when the stratosphere's temperature was relatively warm—Aug. 28 and 21—ozone was very abundant. During the cold days that preceded and followed the warm spell—Aug. 18 and Sept. 9—the ozone layer thinned out. Pioneer ozone researcher Gordon Dobson noticed this phenomenon in the 1950s. He described the ozone hole (or thinning) as largely a dynamic phenomenon with great dependence on stratospheric temperatures and planetary wave patterns.

Source: Adapted from Hiroshi Kanzawa and Sadao Kawaguchi, "Large Stratospheric Sudden Warming in Antarctic Late Winter and Shallow Ozone Hole in 1988," *Geophysical Research Letters* Vol. 17, (Jan. 1990), p. 77.

more rigorously than have their counterparts in Britain and the United States, have made another very interesting observation. This concerns the influence on the ozone layer of the quasibiennial oscillation (QBO) of the equatorial lower stratosphere. The quasibiennial oscillation is a sudden shift in wind patterns, governed by solar variations, that occurs every two years. This shift is of major significance for planetary waves, the giant waves in the stratosphere that move masses of air, as ocean waves move seawater.

Figure 5.4 shows the year-to-year variations of October monthly means of total ozone (top panel) and October monthly means of temperature, at 30 millibars of atmospheric pressure (about 22 km altitude). The asterisks denote the easterly wind phase of the quasi-biennial oscillation, when ozone readings are high; the dots represent the westerly wind phase, when ozone readings over Antarctica are low.

Of this phenomenon, Kanzawa and Kawaguchi (1990) write:

> When planetary waves are active as in 1988, Antarctic temperature in the lower stratosphere can become warmer and the Antarctic ozone hole can become shallower in depth and smaller in area. The reason may be as follows. If planetary waves are active, poleward transport of heat and ozone is effective. Moreover, when the Antarctic lower stratosphere become[s] warm in August and September as a result of large sudden warmings as in the 1988 Syowa case, the Polar Stratospheric Clouds (PSCs) cannot exist so that chemical destruction of ozone may be weaker. . . .
>
> The strong activity of planetary waves in 1988 might be due to an anomaly in sea surface temperature in the equatorial Southern Pacific. . . .
>
> The good correlation between temperature and total ozone oscillations within about two weeks . . . indicates that the ozone increase and decrease of this time scale are due to dynamical transport effects by planetary waves since the temperature oscillation is considered to be due to a dynamical phenomenon. . . .
>
> Dynamical processes are also considered to play an essential role in interannual variation of Antarctic total ozone. The good correlation between interannual variation of October total ozone with lower stratospheric temperature in the Antarctic and the quasibiennial oscillation of the equatorial lower stratosphere . . . may mean that under the easterly (westerly) phase of the equatorial QBO, planetary wave activity is strong (weak) to bring about high (low) total ozone and warm (cold) lower stratosphere in the Antarctic in the recent 10 years. [On

Figure 5.4
TRACKING THE OZONE ANOMALY IN ANTARCTICA

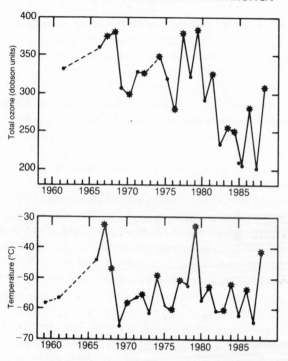

The ozone anomaly at the South Pole can be explained by natural, dynamic causes, without resorting to exotic chemistry concerning man-made CFCs. These data from Japan's Syowa Station in Antarctica show that planetary wave activity is strong enough to penetrate the Antarctic polar vortex and, together with warm stratospheric temperatures, bring about high concentrations of ozone. Yearly variations in the monthly mean values of total ozone (top) and October monthly mean values of temperature (bottom) are plotted for 1960-1985. (Broken lines indicate missing years.)

The asterisks and solid circles denote the easterly wind phase and westerly wind phase, respectively, of the equatorial quasibiennial oscillation (QBO), a phenomenon in which upper atmospheric planetary waves change direction every two years. During the westerly phase of the QBO, planetary wave activity is much stronger. Together with colder temperatures in the stratosphere, this brings about a significant depletion of the ozone layer. Note that ozone concentrations in 1988 were higher than those in 1975, and almost as high as those in 1960.

Source: Adapted from Hiroshi Kanzawa and Sadao Kawaguchi, "Large Stratospheric Sudden Warming in Antarctic Late Winter and Shallow Ozone Hole in 1988," *Geophysical Research Letters*, Vol. 17 (Jan. 1990), p.77.

131

the other hand,] under the westerly phase of the equatorial QBO, planetary wave activity is too weak to bring about low total ozone and cold lower stratosphere in the Antarctic [p. 79].

In conclusion, Kanzawa and Kawaguchi write, "The 1988 phenomenon means that warming and ozone transport by dynamical process can deeply affect the depth and area of the Antarctic ozone hole" (p. 79).

The Japanese are not the only ones looking at the role of the quasibiennial oscillation in ozone fluctuations. James K. Angell, of the National Oceanic and Atmospheric Administration (NOAA) Air Resources Lab published a paper in the September 1990 issue of *Geophysical Research Letters* analyzing the relationship between the stratospheric temperatures above Singapore (which reflect the phase of the quasibiennial oscillation), and the equatorial sea-surface temperatures, and found a close correlation to the thickness of the Antarctic ozone hole. Examining data going back to 1959, Angell found that when the stratospheric temperatures over Singapore increased from one year to the next, ozone depletion in Antarctica was 21 percent greater than in years when the stratospheric temperatures over Singapore decreased. The data also showed that when the June, July, and August equatorial sea-surface temperature increased from one year to the next, Antarctic ozone was depleted 18 percent more than when the equatorial sea-surface temperatures decreased. In the six years during which the stratospheric temperatures at Singapore and equatorial sea-surface temperatures both increased from one year to the next, the values of Antarctic total ozone in the spring have decreased. However, in the six years during which the stratospheric temperatures at Singapore and equatorial sea surface temperatures decreased from one year to the next, the Antarctic total ozone increased.

Based on these observations, Angell made the daring prediction that the 1990 Antarctic ozone hole would be deep. This prediction went contrary to all expectations, because until then there had been a biennial variation in the depth of the Antarctic ozone hole; the hole was deep in odd years and less depleted in even years. Angell was proven correct.

A more recent paper, by Walter Komhyr et al. (1992) in the *Canadian Journal of Physics,* goes even further, documenting the relationships among the equatorial sea-surface temperatures, the quasibiennial oscillation, planetary waves, and the thickness of the ozone layer over Antarctica. The critical point here, seldom mentioned by the proponents of the ozone depletion theory, is the fact that most of the ozone in the world is created over the tropics, where ultraviolet light is most intense and constant. Thus, most of the ozone over the northern latitudes and in Antarctica is transported from the tropical stratosphere to the poles.

The sea-surface temperatures at the equator are critical in creating masses of hot air that rise to the equatorial stratosphere (carrying with them huge amounts of chlorine and bromine compounds from seawater) and push the ozone-rich air toward the pole—a mechanism in which winds, the quasibiennial oscillation, and giant stratospheric planetary waves also play major roles. According to Komhyr, "Because ozone has a lifetime that varies from minutes to hours in the primary ozone region at high altitudes in the tropical stratosphere to months and years in the low stratosphere, changes in atmospheric dynamics have the potential for not only redistributing ozone over the globe, but also changing global ozone abundance."

Komhyr also notes that recorded ozone data in Samoa and Hawaii for the past 14 years show no reduction at all in the thickness of the ozone layer.

Figure 5.5 is a good illustration of the process by which the thickness of the ozone layer is controlled by planetary wave motion. This computer plot of ozone level data provided by NASA's Nimbus 7 satellite over a period of three months in 1989 shows two main waves of ozone moving around the North Pole. The crests of the waves peak at over 475 dobson units; the troughs of the waves are as low as 225 dobson units. Note that the thickness of the ozone layer over Scotland is less than 250 dobson units as a trough passes by on January 30, yet it is more than 475 dobson units barely a month later, as the crest of the wave passes.

The Antarctic 'Reactor Vessel'

One of the most important things to note about the so-called Antarctic ozone hole is that some of the most complex and least-understood atmospheric chemical reactions occur during the four-to-six-week duration of the ozone hole. As mentioned above, the polar vortex seals the Antarctic atmosphere during this period, creating what is essentially an extraordinary chemical reaction

Figure 5.5
HOW PLANETARY WAVE MOTION CONTROLS OZONE LAYER THICKNESS

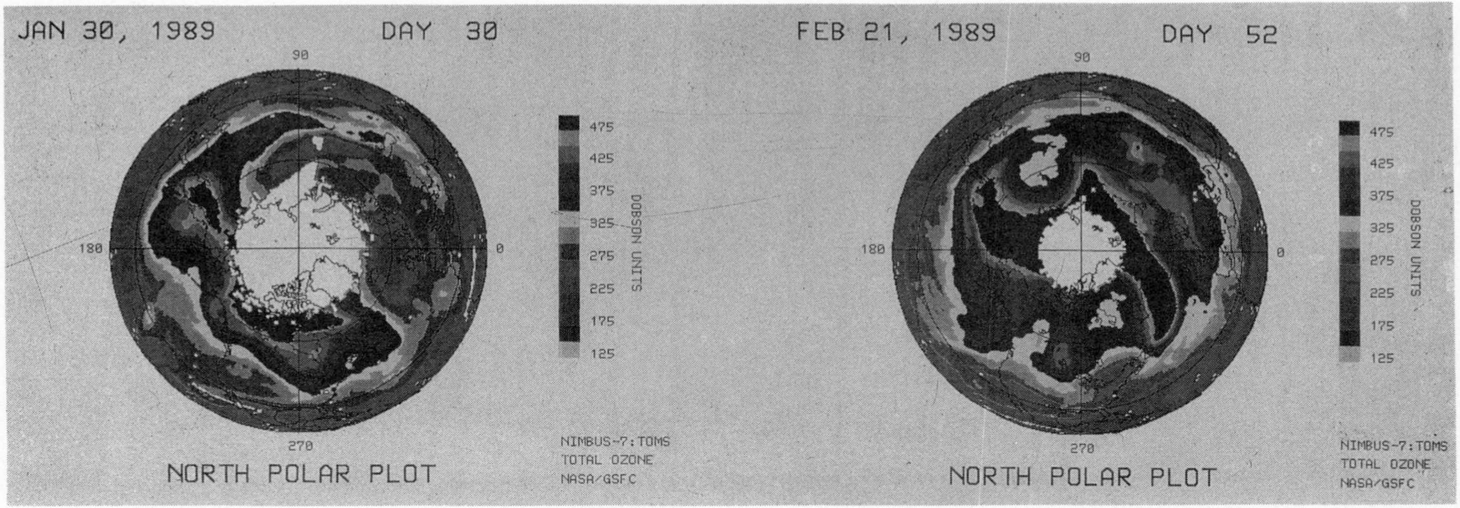
JAN 30, 1989
DAY 30
90
180
0
270
NORTH POLAR PLOT
475
425
375
325
275
225
175
125
DOBSON UNITS
NIMBUS-7:TOMS
TOTAL OZONE
NASA/GSFC
FEB 21, 1989
DAY 52
90
180
0
270
NORTH POLAR PLOT
475
425
375
325
275
225
175
125
DOBSON UNITS
NIMBUS-7:TOMS
TOTAL OZONE
NASA/GSFC

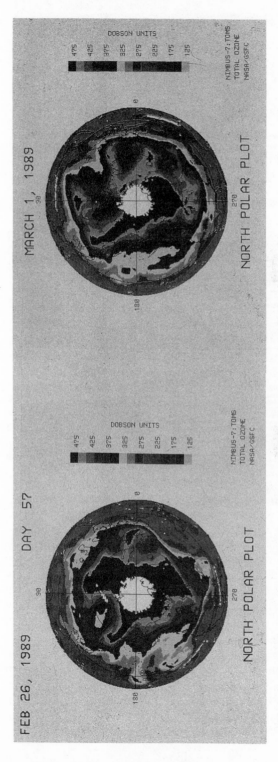

These photographs, generated by computer from data gathered by NASA's Nimbus 7 satellite, show the concentrations of ozone at the North Pole over January, February, and March 1989. The black area represents high concentrations of ozone being carried by planetary wave motion around the pole. The crests of the waves peak at more than 475 dobson units; the troughs of the waves are as low as 225 dobson units. Note that the thickness of the ozone layer over Scotland is less than 250 dobson units, as a trough passes by on Jan. 30, yet it is more than 475 dobson units barely a month later, as the crest of the wave passes.

Figure 5.6
HOW THE CHEMISTRY OF THE STRATOSPHERE
CHANGES AT THE POLAR VORTEX

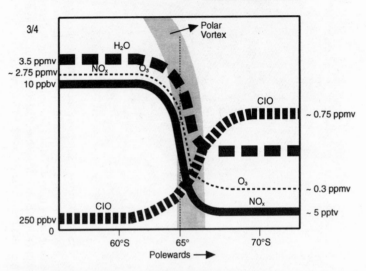

Antarctic Reactor Vessel

Shown are the changes in concentrations of chemicals in the atmosphere as air passes from outside the polar vortex into the so-called Antarctic reactor vessel which, in annual cycles, surrounds and seals off the poles. Note the very sharp drop in concentrations for water vapor, nitrogen oxides, and ozone between the outside of the polar vortex and the inside. Why not campaign to "Save the Earth" from the "nitrogen oxide hole"?

Measurements are in parts per million, billion, or trillion volume.

vessel. As can be seen in Figure 5.6, very dramatic changes occur in the chemical composition of the stratosphere as one flies in the stratosphere from outside the vortex to the vessel inside. The concentrations of many chemicals drop dramatically, including water vapor, nitrogen oxides, and ozone. At the same time, the concentrations of other chemicals, like chlorine monoxide, increase dramatically.

The boundary for these extraordinary changes in chemical concentrations is the wall of the polar vortex. Think of it as a sealed chemical reactor vessel inside which there is a water vapor hole, a nitrogen oxide hole, and an ozone hole—all occurring simultaneously. These chemical conditions exist nowhere else on Earth, except perhaps inside the short-lived Arctic polar vortex. This schematic diagram in Figure 5.6 is presented at scientific meetings, but seldom at public forums or in the news media. Why don't those who fret about the ozone hole also worry about the nitrogen oxide hole, and so on?

Explaining this complex chemistry has been a major problem for the proponents of the ozone depletion theory. From their standpoint, the "discovery" of the so-called ozone hole in 1985 was both a blessing and a curse: a blessing because it revived their sagging fortunes; a curse because F. Sherwood Rowland's theory could not possibly account for the depletion. The latest version of Rowland and Molina's theory predicts a 5 percent depletion of ozone over 100 years. In Antarctica, scientists were observing depletions of 50 percent in a few weeks! Then, a few weeks later, the ozone level was back to "normal" again.

It took two years for the ozone depletion propagandists to come up with an explanation for this anomalous situation.* Mario Molina devised an unbelievably complex chemical theory called "hetero-geneous" or "dimer" chemistry (Molina and Molina 1987). The theory requires very cold temperatures, below $-78°C$, which occur in the Antarctic stratosphere only a few weeks of the year. It also requires the formation of polar stratospheric clouds, which are made up of nitric acid, instead of the water that makes up normal clouds. Finally, Molina's new theory requires sunlight at just the right time.

These conditions can occur in Antarctica only after three to four months of complete darkness enable the stratosphere to cool down to $-78°C$. Then, at the very moment that spring returns and sunlight strikes Antarctica, at that moment, all conditions being right, the stratosphere being primed, the sunlight supposedly sets off a series of very complex reactions that break up the molecules in which chlorine is bound, freeing individual chlorine atoms to destroy the ozone layer.

Molina's chemical formulas are as follows:

* The depletion theorists are not fazed by their own contradictory pronouncements. An article by Martyn Chipperfield in *Nature* Jan. 24, 1991, for example, triumphantly proclaims in the first sentence, "It is now beyond doubt that stratospheric ozone is being destroyed by chlorine derived from man-made chlorofluorocarbons (CFCs)." In the next paragraph, however, Chipperfield warns that ". . . many quantitative details of the Antarctic ozone depletion remain unexplained. . ." (p. 279).

(1) $ClONO_2 + HCl \xrightarrow{\text{ice}} Cl_2 + HNO_3$
(2) $Cl_2 + h\nu \rightarrow 2\ Cl$
(3) $Cl + O_3 \rightarrow ClO + O_2$
(4) $ClO + ClO + M \rightarrow Cl_2O_2 + M$
(5) $Cl_2O_2 + h\nu \rightarrow Cl + ClOO$
(6) $ClOO + M \rightarrow Cl + O_2 + M.$

The net result of this series of complex chemical reactions is that two ozone molecules (O_3) will be turned into three oxygen (O_2) molecules. This is the heart of the explanation that CFCs are depleting ozone in Antarctica.

Let's look at Molina's dimer chemistry in more detail. First, notice that CFCs are not involved in this chemical reaction; the chlorine comes instead from two "reservoirs," $ClONO_2$ and HCl.

Second, ice is needed to begin the reaction, which is why the polar stratospheric clouds are required. This ice is found only when temperatures are colder than $-78°C$ and at an altitude of between 12 and 20 km.

Third, M (a third body) is brought in as a "collisional chaperone" for N_2 and O_2.

Fourth, without sunlight ($h\nu$ stands for a photon of light) this reaction could not occur.

Let's concentrate on Equation (5). This crucial equation says that when the molecule Cl_2O_2 (chlorine peroxide) is struck by sunlight, it will break up into a Cl atom, which goes on to destroy ozone molecules, and ClOO. The ClOO then is presumed to undergo a molecular collision, to give up molecular oxygen and a free chlorine atom. The crucial question is: Given that this theoretical mechanism has never been definitely established in the laboratory, does the chemistry work like this in the stratosphere?

"No," says Igor J. Eberstein of NASA's Goddard Space Flight Center. In a paper published in *Geophysical Research Letters* in May 1990, Eberstein demonstrates that the most likely path of chlorine peroxide photodissociation is into two ClO radicals; that is, back to the monomer. A secondary path of diassociation is Cl_2 and atomic oxygen.

The ozone depletion theorists conveniently ignore these least-energy pathways. They claim that the chemical reaction goes this way:

$$Cl_2O_2 + h\nu \rightarrow Cl + ClOO$$

Eberstein shows that the reaction actually follows one of these two most probable least-energy pathways:

Path (1): Cl_2O_2 + H becomes 2 ClO or
Path (2): $Cl_2O_2 + h\nu \rightarrow Cl_2O + O$

"There is no proven chemical mechanism to account for the creation of the ozone hole," Eberstein said in an interview with one of the authors. "This is a very serious failure. If you have a theory, you should be able to provide a definitive mechanism. Otherwise it is speculation. This Antarctic ozone depletion issue has to be put on a more solid scientific basis."

Eberstein is not alone in criticizing the chemical hocus pocus. Writing in the *Journal of Geophysical Research* on Oct. 20, 1990, W.G. Lawrence and his associates demolish a popular version of Molina's Equation (6) and the presumed gas-phase photodissociation of chlorine dioxide to free up chlorine. After a series of very complex experiments in the laboratory, Lawrence, Clemitshaw, and Apkarian (1990) conclude:

In the spectral range in which it has recently been reported that OClO undergoes unimolecular dissociation to produce Cl + O_2 . . . we have conducted studies to establish that if indeed such a photodissociation channel exists, then its quantum yield is less than 5×10^{-4}. *Such a small quantum yield process would render the photochemistry of OClO irrelevant to the destruction of stratospheric ozone* [emphasis added] [p. 595].

Sunlight is another required element in Molina's "dimer" chemistry. Sunlight is the "trigger" for the chemical reaction that destroys ozone molecules; this is why the ozone hole appears only at the beginning of the Antarctic spring, although the chlorine molecules have been there all through the winter darkness.

Again, reality intrudes. The National Oceanic and Atmospheric Administration (NOAA) announced in September 1990 that its polar satellites were detecting the development of the ozone hole a full month *before* the appearance of sunlight. In other words, the hole is already well developed before sunlight strikes Antarctica, exactly the opposite of what Molina's heterogeneous chemistry theory predicts. If chemical reactions are creating the hole, these reactions are occurring in the darkness, which invalidates the theory.

Not surprisingly, the news media ignored the importance of the NOAA discovery in refuting Molina's dimer chemistry. Instead, the press played the news as another scare story, reporting that the NOAA satellite data showed Antarctic ozone depletion to be more serious than originally thought, because the hole was—unexpectedly—appearing early.

Antarctic Chlorine Comes from Mt. Erebus, Not CFCs

In Molina's dimer chemistry, there are no CFCs. Instead, there are two "reservoir" species of chlorine that are assumed to come from CFCs that have been broken up by ultraviolet radiation elsewhere.

According to the statements of the ozone depletion theorists, the amount of chlorine in the stratosphere has increased nearly fourfold in the past couple of decades, an increase attributable to the breakup of CFCs in the stratosphere. Furthermore, we are ominously told, the levels of chlorine oxide in the Antarctic stratosphere are 1,000 times greater than the natural background levels. (Bear in mind that some species of chlorine in the stratosphere are never measured and those that are, are measured very infrequently and in only a few places.)

This suggests a problem for the CFC ozone-hole theorists. They claim a fourfold increase in total chlorine in the stratosphere from the approximately 0.6 ppb "believed to be" the natural abundance a few decades ago. However, the World Metereological Organization's "Report on Atmospheric Ozone" in 1985 reported that the total amount of chlorine in the stratosphere

is determined largely by the amount of organically bound chlorine in the troposphere, whose background concentration is estimated to be in the range of 2.4–2.8 ppbv chlorine. Hence an equivalent mixing ratio for total chlorine is expected in the stratosphere. Berg et al. (1980) have measured total halogens at ~ 20 km altitude and at various latitudes, using activated charcoal and neutron activation analysis, obtaining values ranging between 2.7 ±0.9 and 3.2 ±0.7 ppb chlorine... [p. 646].

On the one hand, by using high-level research aircraft penetrations of the polar vortex, they have found 100- to 1,000-fold increases in the active forms of chlorine, presumably the result of removal of inhibiting nitrogen locked up in the nitric acid in falling ice crystals from stratospheric clouds. On the other hand, they presume a fourfold increase in total chlorine in the stratosphere as a result of chlorine released from photolytically decomposed CFCs. Even if stratospheric total chlorine has been increased by ultraviolet decomposition of CFCs, the 1,000-fold concentration increases in active chlorine caused by a mechanism depending on temperatures and water concentrations appear to be far more determinative. This is particularly so when it is recalled that an ozone hole was observed in 1958 and that the present ozone hole went from undetected in 1977 to a maximum possible effect in 1987, during which time stratospheric chlorine from CFCs could have increased by not more than 40 percent, even if the claimed man-induced fourfold increase were valid.

As stated by Susan Solomon, in a 1990 article in *Nature* magazine, model calculations "fall short of accounting for the decadal trend in Antarctic ozone when using only chemical processes and assuming that the extent and duration of cloud chemistry in previous years was identical to that of 1987" (p. 352).

As to the source of this enormous amount of chlorine, there is no complicated mystery. It clearly comes from Mt. Erebus, which as noted already, erupts more than 1,000 tons of chlorine a day into the Antarctic atmosphere. Mt. Erebus also knocks down one of the recent claims of the ozone depletion propagandists. Examination of ice core data from Antarctica revealed an increase in chlorine in the snow pack over the past couple of decades. The ozone depletion theorists argued that this chlorine was from CFCs and that it proved their theory.

However, a scientific paper published in *Geophysical Research Letters* in November 1990 refutes this claim. In this paper, Philip Kyle and colleagues Kimberley Meeker and David Finnegan, from the New Mexico Institute of Mining and Technology, state that *all* the excess chlorine found in the Antarctic snow pack may come solely from the active volcano. The scientists examined all the available scientific data on the volcanic emissions between 1972

141

and 1987, concluding that when accurate measurements of the volcanic gases were taken in 1983, the hydrogen chloride (HCl) and hydrogen fluoride (HF) emissions of Mt. Erebus were 1,230 tons per day and 480 tons per day, respectively. These emissions "are extremely high and comparable to the lower limits of total global volcanic emissions," Kyle et al. say. Examining the transport of the chlorine emissions from Mt. Erebus throughout Antarctica, they conclude, "Mount Erebus must be recognized as an important potential and possibly past source of aerosols and could be responsible for inorganic chlorine found in snow and ice from central Antarctica."

Ozone Hole at 'Maximum' in 1987

The final pillar of the dimer theory is knocked down by atmospheric scientist Hugh Ellsaesser, who points out in a recent speech (Ellsaesser 1990):

The ozone hole . . . reflects a process which can occur only in those portions of the atmosphere which are maintained at temperatures below about −80°C (−112°F) for two to three months, during at least the latter half of which, they must also be exposed to sunlight. Such temperatures occur only in restricted vertical layers, roughly 12 to 20 km, within the polar vortices which develop due to radiative cooling when sunlight is absent over the pole in winter, and at the tropical tropopause.

In 1987 the level of ozone within this cold layer over Antarctica fell essentially to zero—less than 5 percent of normal. In other words, the *maximum possible ozone hole occurred in 1987*. The phenomenon does not occur to any appreciable extent over the North Pole because the north polar vortex breaks up and rewarms about the same time as the Sun comes up there. Also, ozone loss is unlikely to ever be detected at the tropical tropopause both because there is little ozone there to be destroyed and because the air there is constantly being flushed out by a slow updraft through the tropical tropopause. Thus, unless there are changes other than the simple addition of more chlorine to the stratosphere, the ozone hole does not appear likely to become any more important than it was in 1987. It should also be noted, that the ozone hole merely causes ultraviolet fluxes at the surface over Antarctica in spring, comparable to what is experienced there every summer.

In other words, the so-called ozone hole is a self-limiting process, which already reached its maximum possible size and intensity in 1987 without producing harm to anyone or anything. Despite all the doomsday scenarios of a monster ozone hole, adding additional chlorine to the stratosphere will not change anything except possibly to make the hole appear a little bit more rapidly every spring.

Does the Sun Influence Antarctica's Ozone Layer?

In Chapter 3, we saw the graph prepared by James Angell showing the close correlation between the thickness of the ozone layer and the number of solar sunspots (Figure 3.2). It is accepted by the scientific community that, indeed, the greatest influence determining the global thickness of the ozone layer is the Sun, through solar radiation. Curiously, however, when it comes to the Antarctic ozone hole, the influence of the Sun is completely ignored or worse; many proponents of the ozone depletion theory become hysterical when any such relationship is proposed. The fact is that the Sun may be the most important factor in the processes that occur inside the polar vortex.

Man's near-space environment is dominated by energy from the Sun, which arrives in the vicinity of the Earth in two forms: as electromagnetic radiation and as the plasma flow of the solar wind. This energy interacts with the Earth's magnetosphere and atmosphere in a variety of ways. Most of the solar electromagnetic energy lies in the visible region of the spectrum and passes directly through the upper atmosphere as sunlight. However, part of the solar energy, in the ultraviolet range, interacts strongly with the upper atmosphere and produces the ionosphere and ozone layers.

The second important carrier of solar energy is the solar wind. The rapid flow of this magnetized plasma against the Earth's magnetic field sets hot plasma and magnetic flux in motion around the Earth. This solar wind has a typical velocity of about 1 million miles (1,610,000 km) per hour, but it can be much faster when the Sun is active. At Earth's orbit, this plasma generally consists of

143

five to ten electrons and protons and one helium ion per cubic centimeter.

As the solar wind approaches the Earth, it intersects an obstacle in its path: our planet and its magnetic field. If the plasma were moving slower than the speed of sound, it would have time to adjust its motion around us—like the water flowing past a ship. However, the wind passes by us supersonically and a shock wave is created, just as one builds up around a supersonic airplane. This bow shock forms some 14 to 16 Earth radii from our planet, facing the Sun. Its location is highly variable because it depends on the wind's speed, which changes with the level of solar activity. The plasma flows across the bow shock into a region called the magnetosheath, which is quite turbulent. Figure 5.7 is a schematic diagram of the structure created by this interaction between the solar wind and the Earth's magnetic field, showing the movement of the particles.

Sunspots, and solar flares associated with sunspots, release large bursts of energy. The Sun normally emits the solar wind at about 1 million miles an hour, but the flares create waves in the wind that are like a tsunami. When these hit the Earth's magnetic field, the field deforms, and as the waves move they generate currents.

Increases in solar activity are manifest in a wide range of effects, including greater numbers of sunspots and solar flares, variations in the solar constant (the amount of energy emitted by the Sun), greater fluxes of solar ultraviolet and extreme ultraviolet radiation, solar X-rays and protons, and geomagnetic storms in the Earth's atmosphere and the Earth's magnetic field.

The solar wind interacts in a very complex way with the Earth's magnetic environment. The most important point, however, as seen in Figure 5.7 and in greater detail in Figure 5.8, is that the Earth has two structures that look like whirlpools, through which the solar and cosmic winds are channeled down to the Earth. These are the only two spots on Earth where this happens, and their location—the poles—is determined by the magnetic field of the Earth. These two whirlpools, or vortex structures, also channel particles from the Earth into outer space. Spacecraft flying down the magnetotail of the Earth have detected large quantities of particles characteristic of life (like oxygen) tens of millions of miles down the magnetotail.

If one follows the vortex down from the magnetosphere, one sees the vortex ringed with a series of structures: first a polar

FIGURE 5.7
THE SOLAR WIND AND THE EARTH'S MAGNETIC FIELD

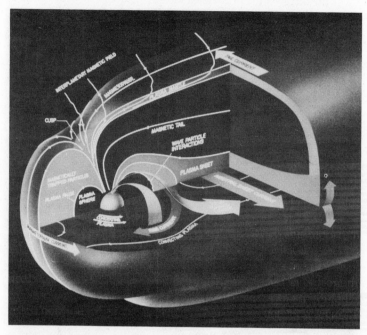

The solar wind, traveling at more than a million miles per hour, collides against the Earth's magnetic field in the area known as the magnetosphere. The magnetosphere acts as a shield, protecting life on Earth from this intense bombardment of energetic solar particles. Some of these particles, however, are trapped in this region and are eventually precipitated into the upper atmosphere near the poles, creating the aurorae and other electromagnetic phenomena.

Source: Courtesy of R.A. Hoffman, NASA Goddard Space Flight Center

electrojet, then the aurora, then the polar jet stream. In other words, the Antarctic ozone hole is contained in its entirety within the vortex structure that starts in outer space. This is one of the most fascinating areas of scientific study today, and clearly further work is needed.

The idea that high-energy protons from the solar wind and flares, channeled by this vortex structure, would play a major role in Antarctic ozone was put forward by physicist Larry Neubauer at

FIGURE 5.8
THE SOLAR WIND AND MAGNETIC FIELDS

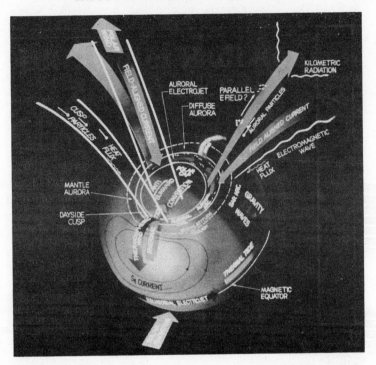

A schematic diagram showing the complex interactions between the solar wind, the interplanetary magnetic field, and the Earth's own magnetic field. Just as some of the most energetic particles of the solar wind enter the Earth's atmosphere at this point, many of the Earth's own particles (some of which are characteristic of life), are transported into outer space and can be found millions of miles away from the Earth in the magnetotail.

Source: Courtesy of R.A. Hoffman, NASA Goodard Space Flight Center

the Second International Symposium on Solar-Terrestrial Influences on Weather and Climate, held in Boulder, Colorado, Aug. 2–6, 1982. Neubauer (1983) made a connection between depletion of ozone and the movement of the jet streams; his work was one of the first theories on the localized destruction of ozone that emphasized the role of the Sun.

According to Neubauer, "As an interplanetary shock wave strikes the Earth's magnetosphere, it injects energetic solar wind particles into the Earth's atmosphere in the 45°–60° latitude band, especially in the 100E–140E longitude region." This, he says, "leads to a sudden increase in ozone-destroying chemical species in the upper stratosphere, caused by both the penetration of the energetic protons into the thermosphere, increasing NO, which reduces the dissociation rate at lower altitudes, causing a greater increase in NO in the mesosphere and the stratosphere; and the deep penetration of energetic protons to the lower-altitude upper stratosphere." The result, Neubauer says, is a sudden drop in ozone above the poles.

Neubauer's theory was confirmed recently by observations made by South African space physicists Judy Stephenson and Malcolm Scourfield. Writing in *Nature* July 11, 1991, they state: "We examine the depletion of stratospheric ozone caused by the reaction of ozone with nitric oxide generated by energetic solar protons, associated with solar flares. During large solar flares in March 1989, satellite observations indicated that total column ozone was depleted by 19% over 120% of the total area between the South Pole and latitude 70°S." The authors conclude that "the influence of SPEs [solar proton events] should not be ignored in any assessment of the Earth's ozone budget."

Linwood B. Callis and coworkers reported in the February 1991 issue of the *Journal of Geophysical Research* that their model could account for 73 percent of the decline in globally averaged total ozone from the solar maximum of 1979 to the solar minimum of 1986 as caused by decreasing solar ultraviolet (20 percent) and to the NO_x generated by the unusually intense solar REP (relativistic solar electron precipitation) events (56 percent) during this half solar cycle.

A group of Italian scientists, led by physicist Giovanni Moreno of the Universita La Sapienza in Rome, Italy, has a different theory. They think that the energetic charged particles that create the aurora have enough energy to break up the ozone molecules inside the polar vortex in Antarctica. According to Moreno's group (de Petris et al.):

Most theories so far proposed to explain the [ozone hole] are focused on chemical and dynamical processes occurring in the atmosphere. Less attention has been paid to the effects of the solar activity. The main argument used to exclude this possibility has been the lack of a periodicity in the ozone trend similar to that of the 11-year sunspot cycle. It is well known, however, that the sunspot number is not the best parameter to describe the terrestrial effects of the solar evolving features. Other solar phenomena, having an impact on the terrestrial environment, may, on the other hand, induce variations over time scales much longer than 11 years.

Solar wind macrostructures overtaking the Earth have a strong influence on several terrestrial phenomena. Space observations have also shown that high- and low-speed solar wind streams have different origins. Two classes of high-speed streams were identified in the interplanetary medium: (i) regular and recurrent streams coming from coronal holes, (ii) complex and transient streams associated with energetic flares. A third class of streams, generally characterized by low-speed and short duration, seems to relate to coronal mass ejections, eruptive prominences or disappearing filaments.

Following a suggestion given in a previous report, here we explore the possibility that the solar wind affects the atmospheric ozone equilibrium in the polar regions through aurorae. Aurorae are a manifestation of a large-scale electrical discharge process surrounding the Earth, powered by a dynamo produced by the solar wind-magnetosphere interaction. A rough order of magnitude estimate of the aurora energy content suggests, indeed, that these phenomena could have eventually relevant effects on the atmospheric ozone. . . .

The authors conclude, "We stress again that the present experimental evidence allows us to consider positively that geomagnetic phenomena, such as aurorae, induced by the solar-related disturbances, may contribute significantly to the observed decline of the atmospheric ozone level over the Antarctica."

There is a third theory, proposed by Tom Valentine in *Magnets* magazine. Citing the work of Michael Faraday on magnetism, Valentine argues that the Antarctic ozone hole is the result of the geomagnetic field of the Earth. In his experiments on the effect of magnetism on gases, Faraday discovered that magnetism had a very great influence on many gases, especially oxygen, and little influence on some others, such as nitrogen. Paramagnetic molecules like oxygen, Faraday demonstrated, are aligned by the magnetic field lines, and their concentrations can be determined by the intensity of the magnetic field and atmospheric temperatures; the lower the temperatures, the more paramagnetic oxygen, and thus ozone, becomes.

This chapter has presented many different theories that explain different aspects of the Antarctic ozone hole. The electromagnetic and atmospheric phenomena occurring at the poles are very complex, and the importance of this chapter does not lie in proving that any of these alternative ozone-depletion theories is correct, but in showing that this is a fascinating subject that has to be studied in much greater detail. What is certain, however, is that the role of CFCs in the Antarctic ozone hole is far from established.

References

James K. Angell, 1989. "On the Relation between Atmospheric Ozone and Sunspot Number," *Journal of Climate,* Vol. 2 (November), pp. 1404–1415.

————, 1990. "Influence of Equatorial QBO and SST on Polar Total Ozone, and the 1990 Antarctic Ozone Hole," *Geophysical Research Letters,* Vol. 17, No. 10 (September) pp. 1569–1572.

W.W. Berg, et al., 1980. "First Measurements of Total Chlorine and Bromine in the Lower Stratosphere," *Geophysical Research Letters,* Vol. 7, pp. 937–940.

Linwood B. Callis, D.N. Baker, J.D. Blocke, et al., 1991. "Precipitating Relativistic Electrons: Their Long-term Effect on Stratospheric Odd Nitrogen Levels," *Journal of Geophysical Research,* Vol. 96 (Feb. 20), pp. 2969–2976.

Martyn Chipperfield, 1991. "Stratospheric Ozone Depletion Over the Arctic," *Nature,* (Jan. 24), pp. 279–280.

Shigeru Chubachi, 1985. "A Special Ozone Observation at Syowa Station, Antarctica from February 1982 to January 1983." In *Atmospheric Ozone: Proceedings of the Quadrennial Ozone Symposium held in Halkidiki, Greece, Sept. 3–7, 1984.* Boston: D. Reidel Publishing Company.

Shigeru Chubachi and Ryoichi Kajiwara, 1986. "Total Ozone Variations at Syowa," *Geophysical Research Letters,* Vol. 13 (November Supplement), pp. 1197–1198.

M. de Petris, Giovanni Moreno, M. Gervasi, et al., 1990. "Interplanetary Perturbation-induced Effects on Polar Ozone Level," *Annales Geophysicae* (March).

Gordon M.B. Dobson, 1968. "Forty Years Research on Atmospheric Ozone at Oxford University: A History," *Applied Optics,* Vol. 7, No. 3, pp. 387–405.

Igor J. Eberstein, 1990. "Photodissociation of Cl_2O_2 in the Spring Antarctic Lower Stratosphere," *Geophysical Research Letters,* Vol. 17 (May), No. 6, pp. 721–724.

Hugh W. Ellsaesser, 1990. "Planet Earth: Are Scientists Undertakers or Caretakers?" Keynote Address to the National Council of State Garden Clubs meeting, Hot Springs, Arkansas, Oct. 7.

J. C. Farman, B.C. Gardiner, and J.D. Shanklin, 1985. "Large Losses of Total Ozone in Antarctica Reveal ClO_x/NO_x Interaction," *Nature,* Vol. 315 (May 16), pp. 207–210.

R. A. Hoffman, 1988. "The Magnetosphere, Ionosphere and Atmosphere as a System: Dynamics Explorer 5 Years Later," *Reviews of Geophysics,* Vol. 26, No. 2 (May), pp. 209–214.

D. Hoffman and J. Rosen, 1985. "Antarctic Observations of Stratospheric Aerosol and High Altitude Condensation Nuclei Following the El Chichón Eruption," *Geophysical Research Letters,* Vol. 12, pp. 13–16.

Hiroshi Kanzawa and Sadao Kawaguchi, 1990. "Large Stratospheric Sudden Warming in Antarctic Late Winter and Shallow Ozone Hole in 1988," *Geophysical Research Letters,* Vol. 17 (Jan.), pp.77–80.

W.D. Komhyr, et. al, 1992. "Possible Influence of Long-Term Sea Surface Temperature Anomalies in the Tropical Pacific on Global Ozone," *Canadian Journal of Physics* (in press).

Philip R. Kyle, Kimberley Meeker, and David Finnegan, 1990. "Emission Rates of Sulfur Dioxide, Trace Gases, and Metals from Mt. Erebus, Antarctica," *Geophysical Research Letters,* Vol. 17, No. 12, pp. 2125–2128.

W.G. Lawrence, K.C. Clemitshaw, and V.A. Apkarian, 1990. "On the Relevance of OClO Photodissociation to the Destruction of Stratospheric Ozone," *Journal of Geophysical Research,* Vol. 95, No. D11 (Oct. 20), pp. 18,591–18,595.

L.T. Molina and M.J. Molina, 1987. "Production of Cl_2O_2 from the Self-Reaction of the ClO Radical, *Journal of Physical Chemistry,* Vol. 91, pp. 433–436.

Larry Neubauer, 1983. "The Sun-Weather Connection—Sudden Stratospheric Warmings Correlated with Sudden Commencements and Solar Proton Events." In *Weather and Climate Responses to Solar Variations.* Ed. Billy M. McCormac. Boulder: Colorado Associated University Press.

———, (undated). "A Possible Sun-Weather Mechanism." Unpublished paper. (Physics Department, Brigham Young University, Provo, Utah).

P. Rigaud and B. Leroy, 1990. "Presumptive Evidence for a Low Value of Total Ozone Content above Antarctica in September, 1985," *Annales Geophysicae,* Vol. 11, pp. 791–794.

Susan Solomon, 1990. "Progress Towards a Quantitative Understanding of Antarctic Ozone Depletion," *Nature,* Vol. 347, pp. 347–354.

Judy Stephenson and Malcolm Scourfield, 1991. "Importance of Energetic Solar Protons in Ozone Depletion," *Nature* (July 11), pp. 137–139.

Tom Valentine, 1987. "Magnetics May Hold Key to Ozone Layer Problems: Going Back to Michael Faraday for Answers," *Magnets in Your Future* (published quarterly by L.H. Publishing Agency, P.O. Box 580, Temecula, Calif. 92390.)

World Meterological Organization, 1985. *Global Ozone Research and Monitoring Project Report No. 16.* 3 vols. Geneva: WMO.

6

Ultraviolet Radiation: Friend or Foe?

The ozone depletion scare rests on the assertion that CFCs, by depleting the ozone layer, will allow an increase in the amount of "harmful" ultraviolet radiation reaching the ground, thus causing a cataclysmic increase in human skin cancer rates. If the ozone scare were not tied to this environmental factor, it would not even be an issue today. This chapter examines whether ultraviolet light is as harmful and dangerous as the environmentalists claim it to be.

All life on Earth is dependent upon the Sun, the source of all light. Without the Sun, the Earth would be a frozen ball in space.

The Earth's biosphere is sustained by the energy in sunlight. The Sun provides the electromagnetic energy for plants to carry on photosynthesis, and this process transforms it to energy that is stored in plants in the form of carbohydrates, proteins, and fats, to be transferred to animals and humans upon consumption. Similarly, the Sun provides the energy necessary for the growth of trees, and it is this energy that is released, in the form of heat, upon combustion of the wood. Fossil fuels are an even more compact form of stored photosynthetic energy from a previous era of plant life.

Although most people associate sunlight with visible light, the

Figure 6.1(a)
SPECTRAL CHARACTERISTICS OF THE SUN'S RADIATION

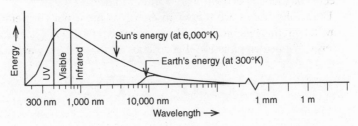

Figure 6.1(b)
AMOUNT OF SUN'S RADIATION TRANSMITTED THROUGH THE ATMOSPHERE

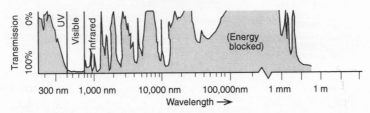

electromagnetic spectrum emitted by the Sun and other stars is much broader: It ranges from cosmic rays, with a wavelength of 0.00001 nanometers (a nanometer, nm, is 1 billionth of a meter) to about 4,990 kilometers (3,100 miles) for waves. A multilayered atmosphere surrounds the Earth, filtering and modulating the electromagnetic waves, such that most of this electromagnetic spectrum is filtered out before the rays reach the Earth.

Figure 6.1 shows how the atmosphere's filtering mechanism works. Figure 6.1(a) shows the spectral distribution of the energy emitted by the Sun, while the shaded area in Figure 6.1(b) shows the energy absorbed by the Earth's atmosphere. There are certain "windows" that allow as much as 100 percent of the Sun's energy to go through the atmosphere. Notice the similarity in the shapes of the emission curve from the Sun and the Earth's transmission curve. Most interesting, the electromagnetic waves in the visible range coincide with an atmospheric window and the peak level of energy from the Sun. Clearly, the Earth is "tuned" to the energy of the Sun.

The visible part of the electromagnetic spectrum is the very

152

narrow range that humans perceive, and accounts for less than 1 percent of the energy in the spectrum. Figure 6.2 shows the electromagnetic spectrum from X-rays to infrared waves.

Ultraviolet waves are higher in energy content but shorter in wavelength than visible light. They comprise 5 percent of the total energy from solar radiation that reaches Earth. Infrared waves, at the opposite end of the visible spectrum, are lower in energy content and longer in wavelength than visible light. They comprise 40 percent of the solar radiation energy that reaches Earth. Both ultraviolet and infrared waves are invisible, and both penetrate and affect the body. Infrared waves are felt as heat; ultraviolet waves are not felt under normal conditions.

The ultraviolet spectrum spans the equivalent of more than three octaves: from 400 nm to roughly 40 nm. To understand how this broad range of ultraviolet light affects the human body, it is convenient to subdivide it into three separate bands: ultraviolet A, ultraviolet B, and ultraviolet C. The longest-wavelength band is ultraviolet A (400 to 320 nm). As can be seen in Figure 6.2, ultraviolet A is further subdivided into longwave and shortwave ultraviolet A. Ultraviolet B (320 to 286 nm) is known as the biologically active range of the ultraviolet spectrum because it has the most obvious effect on living matter. The ultraviolet radiation with the shortest wavelength is ultraviolet C (286 to 40 nm).

Ultraviolet C, the most energy-dense of ultraviolet light, reaches the surface of the Earth in only trace amounts. The rest of the ultraviolet C is filtered out by oxygen molecules (O_2) in the atmosphere. The energetic photons in ultraviolet C split the oxygen molecule, producing two oxygen atoms. This process requires energy, which comes as a result of the higher-energy, shorter-wavelength ultraviolet C being turned into less-energetic, longer-wavelength light. The oxygen atoms are then free to recombine into oxygen or into ozone molecules (which each contain three oxygen atoms, O_3). By this process, billions of tons of ozone molecules are created every second.

Ozone molecules also filter out ultraviolet radiation, starting at the ultraviolet C range all the way through the ultraviolet B spectrum. Like oxygen molecules, ozone molecules are destroyed almost as soon as they are created, by the action of ultraviolet light. The end result, however, is that oxygen and ozone filter out almost all ultraviolet C and most ultraviolet B before these electromagnetic waves reach the surface of the Earth. Interestingly, however,

Figure 6.2
ELECTROMAGNETIC SPECTRUM IN THE ULTRAVIOLET AND VISIBLE RANGE

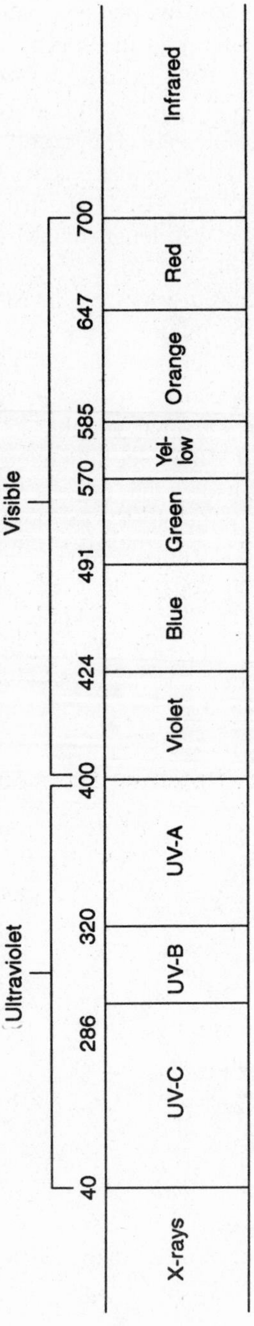

154

oxygen and ozone filter out very little ultraviolet A, which reaches the ground almost in full strength, just like light in the visible spectrum. The amount of ultraviolet A that reaches the ground is 100 times greater than the amount of ultraviolet B that reaches the ground.

Because ozone filters out virtually no ultraviolet A, the threat represented by a depletion of the ozone layer concerns itself exclusively with the increase in the ultraviolet B radiation spectrum. Since ultraviolet B causes sunburn in human beings, it is *assumed* that it also causes cancer. However, although the epidemiological evidence indicates that ultraviolet B may be responsible for a large percentage of benign skin cancers, the evidence does not indicate that it is responsible for the malign types of skin cancers, the ones that kill people. According to the American Academy of Dermatology, malignant melanoma occurs almost always in body areas not normally exposed to solar rays—between the toes, on the soles of the feet, between fingers and on the palms of the hands, under arms, and on the buttocks. It is noteworthy that brown-skinned and black-skinned people rarely get easily treatable skin cancer, but are susceptible to malignant melanoma.

Ozone depletion theorists tend to lump all types of skin cancer together as malign.

Is Ultraviolet Radiation Increasing or Decreasing?

As noted in Chapter 3, the Ozone Trend Panel's Robert Watson announced on March 15, 1988, that the ozone layer above the United States and Europe had been depleted. According to Watson, the ozone layer had decreased more than 3 percent in northern latitudes between 1969 and 1986. The announcement, which was heralded with newspaper headlines about expected new epidemics of skin cancer, had the expected result of generating public hysteria.

That there was no scientific evidence presented of such depletion is documented in Chapter 3. Even if we assume, for argument's sake, however, that there has been a 3 percent depletion of the ozone layer as claimed, what does this actually mean? According to Robert Watson's claims and F. Sherwood Rowland's official CFC depletion theory, a 1 percent decrease in stratospheric ozone will cause a 2 percent increase in the amount of ultraviolet radiation reaching the surface of the Earth. If the CFC depletion theory were correct, that means that ultraviolet radiation should have increased by 6 percent in the same period of time. Such an enormous rise in ultraviolet radiation should be easily detectable, and should have by now, if the theory were correct, produced visible effects on plant and animal life.

In reality, the data show exactly the opposite of what the Ozone Trends Panel's findings indicate. In a study published in *Science* magazine Feb. 12, 1988, Joseph Scotto, of the Biostatistics Branch of the National Cancer Institute, presents hard scientific evidence that the amount of ultraviolet B radiation reaching ground levels across the United States had *not increased* but, in fact, had significantly *decreased* between 1974 and 1985. The study, which has been ignored by the international news media, was based on readings from a network of ground-level monitoring stations that has been tracking measurements of ultraviolet radiation since 1974.

The Scotto study (Scotto et al. 1988) states:

> Average annual R-B [R-B refers to Robertson-Berger meters, in which UV radiation is measured] counts for two consecutive 6-year periods (1974 to 1979 and 1980 to 1985) show a negative shift at each station, with decreases ranging from 2 to 7 percent.... [Figure 6.3] shows that there are no positive trends in annual R-B counts for 1974 to 1985.... The estimated average annual change varied from −1.1 percent at Minneapolis, Minnesota, to −0.4 percent at Philadelphia, Pennsylvania. For all the stations the R-B counts dropped an average of 0.7 percent per year since 1974... [p. 762].

Scotto then reports that his instruments are confirming what many other individual instruments have been recording: a drop in ultraviolet radiation reaching the Earth:

> These results are consistent with earlier reports that used R-B data for a shorter time period from 1974 to 1979 and Dobson meter total column ozone data for the period from 1970 to 1982. Although recent measurements of stratospheric ozone from satellite instruments indicate that total column ozone is being depleted during the 1980s, anticipated resultant increases in solar ultraviolet B were not evident. . . .
>
> Monthly trends of estimated ultraviolet B levels showed

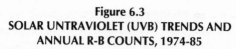

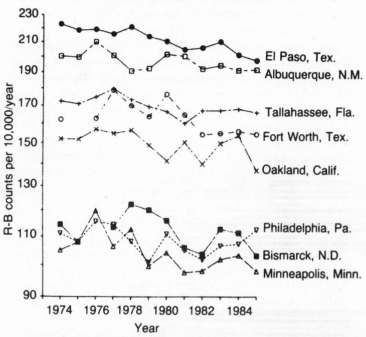

Figure 6.3
SOLAR UNTRAVIOLET (UVB) TRENDS AND
ANNUAL R-B COUNTS, 1974-85

Notice the clear trend toward lower ultraviolet radiation. The greatest total decreases occurred at the field monitoring stations in El Paso, Texas, and Minneapolis, Minnesota.

Source: Scotto et al., "Biologically Effective Ultraviolet Radiation: Surface Measurements in the United States, 1974–1985," *Science*, Feb. 12, 1988

consistent decreases at each field station. The seasons with the greatest relative decreases were the fall (October, November, or December) and winter (January, February, or March). Analysis of peak daily ultraviolet B measures for each of the three 10-day periods within each month showed that the peak day, which is usually cloudless, also had consistent downward trends for the 12-year period. The data were also analyzed with annual calibration factors excluded and the findings remained unchanged.

In a subsequent issue of *Science,* Nov. 25, 1988, Scotto rejects the possibility that urban air pollution was scattering incoming ultraviolet B and thus causing the decrease in ultraviolet B reaching the Earth. Scotto points to data from the air station at Mauna Loa, Hawaii, "which is relatively free of urban air pollution"; yet, he says, "preliminary analysis of data from this site shows no increase in "ultraviolet B radiation from 1974 to 1985."

Exactly what is causing the consistent decrease in ultraviolet radiation at the Earth's surface? Scientists don't really know. It is an interesting question and shows how little is actually known of the dynamics of the biosphere. There are three hypotheses regarding the consistent reduction in ultraviolet radiation reaching the ground:

(1) that the ozone layer is thickening, and thus filtering out more ultraviolet radiation;

(2) that there is more ozone in the troposphere, allegedly the result of urban air pollution, which filters out more ultraviolet radiation; and

(3) that the role of other dynamic processes and compounds in filtering ultraviolet light has been underestimated. Water vapor and clouds, for example, can filter ultraviolet radiation. An increase in water vapor in the air, therefore, would decrease ultraviolet radiation reaching the ground. Likewise, an increase in sulfur dioxide, either from industrial pollution or volcanic eruptions, could be the culprit.

How did the U.S. government and the environmentalists react to the publication of Scotto's National Cancer Institute data? According to Scotto, he has been the victim of an "inquisition." He was unable to extend his study beyond 1985, because funding for most of the ultraviolet recording stations was cut and the stations were shut down. And, although he is a world-renowned cancer expert, Scotto no longer receives funding to travel to international conferences to present his findings.

Is ultraviolet light at the Earth's surface increasing with the depletion of ozone as the environmentalists assert? The data that could answer that question are being suppressed. Environmentalist pressure groups have gone from one agency to another in Washington, attacking the scientists who support the maintenance of the ultraviolet radiation measuring stations. And, although the U.S. government is now spending more than $3.5 billion a year researching "climate change," "ozone depletion," and "global warming," it has been decided that Washington cannot spare a few thousand dollars to keep these monitoring stations operating.

There is other evidence that the level of ultraviolet light reaching the Earth has not increased. Some of this has been brought to light by Stuart Penkett, of the School of Environmental Science at the University of East Anglia, England. Penkett (1989) reports that the Hohenpeissenberg Observatory in Bavaria, Germany, has recorded decreases in ultraviolet radiation reaching the Earth's surface of 0.9 percent at noon and 0.5 percent throughout the day. The measurements were taken between 1968 and 1982. During this period, according to the ozone propagandists, the ozone layer should have decreased 1.5 percent, generating a 3.0 percent increase in the amount of ultraviolet light reaching the Earth's surface.

The ozone hole propagandists also ignore more extensive records of ultraviolet readings taken over the past 20 years. A case in point is data collected for 15 years by the Fraunhofer Institute of Atmospheric Sciences in Germany. In the early 1970s, the institute set up ozone and ultraviolet monitoring devices at several levels in the Bavarian mountains. These instruments have produced a vast body of data. Unfortunately, the institute's founder, Reinholt Reiter, retired in 1985, and his replacement, Wolfgang Seidler, is an enthusiastic proponent of the ozone-depletion and global-warming theories. Seidler has adamantly refused to publish the institute's data, and has gone so far as to stop the researchers who collected it from returning to the institute to use it! Furthermore, in 1990, Seidler began to shut down the monitoring instruments, with the apparent reason that the institute will henceforth rely on computer models instead of observational data.

Geography vs. Hysteria

On an annual mean basis, the amount of ultraviolet radiation reaching the Earth's surface increases about 50-fold, as one travels from the poles toward the equator. Roughly, ultraviolet radiation doubles every 1,000 miles from the poles to the equator, and ultraviolet light reaching the surface of the Earth at the equator is 5,000 percent more intense than at the poles. This is the equivalent of a 1 percent increase every 6 miles approaching the equator. Ultraviolet light also increases with altitude, doubling from sea level to about 15,000 feet—or, roughly, 1 percent per 150 feet

In other words, a 1 percent decrease in ozone, causing a p sumed 2 percent increase in ultraviolet radiation reaching

Earth's surface, is equivalent to a 12-mile displacement toward the equator. Thus, the worst-case scenario predicted by the environmentalists to result from continued use and release of CFCs—depletion of the ozone layer by 5 percent—would be equivalent to about a 60-mile displacement toward the equator. This is like moving from New York City to Philadelphia, or staying at the same latitude but moving to a house located at an altitude 1,500 feet up a mountain.

In the longer term, human populations that have lived for generations in the same geographic area have developed skin types adapted to the exposure to ultraviolet light expected in that latitude. Those living closest to the equator have the highest concentrations of protective melanin (dark skin pigmentation). Those living closer to the poles have the smaller concentrations of melanin. The divine wisdom embodied in this natural variation in skin pigmentation will become clearer below, as we discuss the hazards of too little exposure to ultraviolet light, leading to inadequate synthesis of vitamin D and crippling cases of rickets.

As civilization and lifestyles have advanced in the last several hundred years, human populations have become more mobile, not necessarily staying in the geographic areas to which their genetic skin type is best adapted. The trend in population movement in the past decades in the United States and Europe, for example, has been to go southward, toward the equator and to more sunny climates. Florida, Texas, California, Italy, Spain, and Australia have become favorite vacation resorts and residences. The mobility of the fairer-skinned toward the equator increases the susceptibility to skin cancer of that group.

Does everyone who moves south toward the equator worry that he or she may have increased the daily dose of "harmful" ultraviolet radiation by 100 to 500 percent, depending on where the person lived originally? The Environmental Protection Agency has not issued a single warning that moving from New York to Miami, for example, may cause skin cancer (or, even more relevant, that removal of the smog layer from Los Angeles should increase skin cancer incidence by about 50 percent). Airline advertisements for trips to the Caribbean do not warn travelers from New York, Montreal, or Germany that they may be exposed to a 400 percent increase in ultraviolet radiation.

Given the lack of concern about such massive increases in ultraviolet radiation, why should anyone worry about an alleged 10

Carlos de Hoyos

Would any of these sunlovers seriously worry about skin cancer when they contemplate a move 60 miles nearer to the equator? That move would be the equivalent of the projected 5 percent decrease in the stratospheric ozone layer.

percent increase in ultraviolet radiation 100 years hence because of the continued release of CFCs?

In a 1989 paper, Norwegian scientists Arne Dahlback, Thormod Henriksen, Søren H. H. Larsen, and Knut Stamnes conclude that "depletions of the ozone layer up to about 15 to 20 percent would have a rather small effect on the life on Earth" (p. 624). This level of ozone depletion would be equivalent to increases in ultraviolet radiation of between 30 and 40 percent.

The intensity of ultraviolet light is significantly higher at 40°N than 60°N. It is lowest in the mornings and late afternoon and most intense at noon. Figure 6.4 shows the annual effective ultraviolet dose as a function of the geographical latitude. The dose is set at 100 for Oslo, Norway, and it shows the dramatic increase of the effective ultraviolet dose as one approaches the equator. Figure 6.5 shows their measurements of ozone levels and ultraviolet radiation in Scandinavia for a 12-year period, which indicates no observable trend. "It is quite normal that the ozone layer changes slightly from one year to another," the authors state.

The Norwegian scientists conclude by analyzing the most catastrophic scenarios of ozone depletion. "It appears that a depletion of the ozone layer over Scandinavia of 50 percent (which is far more than all prognoses from atmospheric models) would give

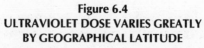

Figure 6.4
ULTRAVIOLET DOSE VARIES GREATLY
BY GEOGRAPHICAL LATITUDE

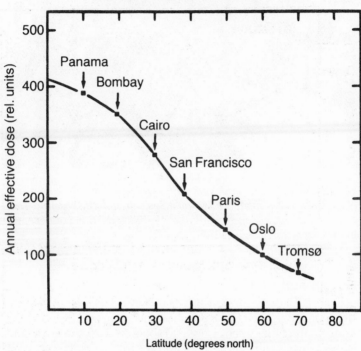

Shown is the annual ultraviolet radiation dose for selected cities at different geographical latitudes in the Northern Hemisphere. Under the worst case scenario of ozone depletion, the increase in the amount of ultraviolet radiation reaching the ground is expected to be 10 percent. Moving from Oslo to Panama represents an increase in ultraviolet exposure of 300 percent, while a move to balmy San Francisco from Oslo is an increase of 100 percent. In fact, a 10 percent increase in so-called harmful ultraviolet radiation, declared to be a disaster by the environmentalists, is the equivalent of moving a mere 60 miles closer to the Equator.

Source: Adapted from Arne Dahlback, Thormod Henriksen, Søren H. Larsen, and Knut Stamnes, "Biological UV-Doses and the Effect of an Ozone Layer Depletion," *Photochemisty and Photobiology, Vol. 49, (1989)*, p. 621.

Figure 6.5
NO OBSERVABLE TREND IN OZONE OR
ULTRAVIOLET LEVELS IN PAST 12 YEARS

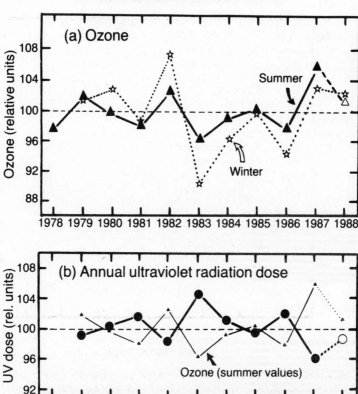

Norwegian measurements demonstrate there is no observable trend, neither increase nor decrease, in ozone (a) or ultraviolet radiation (b) values for the past 12 years. If the ozone depletion theory were correct, ozone values should have gone down more than 3 percent, and ultraviolet radiation values should have therefore increased by more than 7 percent.

Source: Adapted from Arne Dahlback, Thormod Henriksen, Søren H. Larsen, and Knut Stamnes, "Biological UV-Doses and the Effect of an Ozone Layer Depletion," *Photochemisty and Photobiology, Vol. 49, (1989),* p. 621.

these countries an effective ultraviolet dose similar to that normally obtained in California or the Mediterranean countries," they write (p. 624).

In the same paper, the Norwegian scientists also refute the scare story that an ozone hole like the one in Antarctica will appear in the Arctic and gobble up all the ozone in northern latitudes. They write:

> The so-called ozone hole in Antarctica is a transient spring-time depletion of the [Antarctic] ozone layer which is connected to the polar vortex. . . . If we assume a similar depletion over Scandinavia (for example; if we moved the [Antarctic] ozone hole [to the Arctic]) the annual effective ultraviolet-dose would increase by approximately 22 percent. . . . It is of interest to note that one would attain a similar increase in the annual ultraviolet-dose by moving approximately 5° to 6° towards lower latitudes; for example from Oslo to Northern Germany [p. 624].

In plain English, this means that in the real world, the worst doomsday ozone depletion scenario is the equivalent of moving from San Francisco to San Diego, or Chicago to Saint Louis, or New Orleans to Miami.

The environmentalists' comeback is to allege an epidemic of skin cancer in Australia and New Zealand—caused by the ozone hole. Here, the scientific evidence is most interesting. As we have seen already, ultraviolet radiation increases dramatically the closer one moves toward the equator. Although Australia and New Zealand are "down under" in the Southern Hemisphere, they are much closer to the equator than is Great Britain. Thus, a move by fair-skinned Europeans to Australia or New Zealand means an increased ultraviolet exposure of between 250 and 500 percent (Figure 6.6). As a result of this dramatic increase in exposure, there is no question that light-skinned people will have an increase of skin cancers.

The ozone depletion theorists never mention that skin cancer is nearly unknown in dark-skinned and black-skinned individuals, whose skin pigmentation equips them with adequate Sun protection for the tropics. A proper scientific approach to the question of whether ozone depletion, or the so-called ozone hole in Antarctica, has increased skin cancer in Australia and New Zealand would

Figure 6.6
ULTRAVIOLET RADIATION AND SKIN CANCER VARY WITH LATITUDE, SEASON, AND CLIMATE

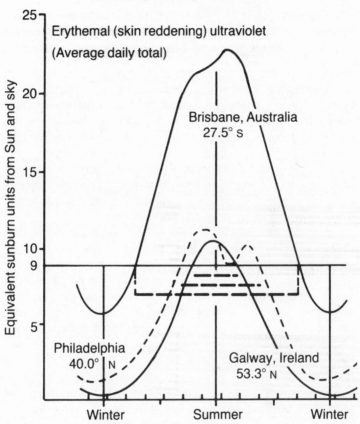

Caucasians living in Australia have high rates of skin cancer because levels of effective ultraviolet radiation in Australia are more than twice those in Philadelphia or England. The environmentalists hysterically allege that "ozone depletion" is causing a skin cancer epidemic in Australia, but the fact is that white-skinned Europeans have settled a continent where the intensity of ultraviolet radiation is 200 to 300 percent greater than in their original lands. By comparison, the predicted 10 percent increase in ultraviolet radiation due to ozone depletion is rather insignificant. Australian aborigines, meanwhile, show no cases of skin cancer, because their dark skins, appropriate for the tropics, effectively filter the ultraviolet radiation.

Source: Adapted from J.D. Everall, "Distribution and General Factors Causing Chronic Actinic Dermatosis," *Research in Photobiology*, Amleto Castellani, ed. (New York: Plenum, 1977).

be a study of the native population. Skin cancer studies have omitted this population, however, since the cases of skin cancer among the aborigines are so rare that their respective governments do not keep records of them.

Vegetation and Ultraviolet Dosage

The ozone depletion theorists also argue that vegetation will be destroyed by increases in ultraviolet radiation. Where is the evidence? Fruits, vegetables, and all kinds of plants, animals, and insects have been transplanted from equatorial latitudes to the northern latitudes, and vice versa, with no observable damage from massive increases in ultraviolet radiation. This "experiment" has been going on for hundreds of years in the New World, as part of the process of European colonization of the American continent. In fact, the scientific evidence demonstrates higher crop yields from higher exposure to ultraviolet radiation.

Ultraviolet radiation, of course, is essential to all plants. The ozone depletion researchers have done their experiments on the subject by giving plants massive overdoses of ultraviolet radiation. They have then used these results to argue that any increase in ultraviolet light would be harmful. The reader can carry out this simple experiment to test the methodology of these researchers: On the next sunny day, take a magnifying glass and focus the rays of the Sun onto a single leaf. Within a very short time, the leaf will begin to burn. Does that mean that the Sun is dangerous to plant life? According to these researchers, the answer is "yes," for this is exactly how they have carried out their experiments to prove that ultraviolet radiation damages plant life.

The Skin Cancer Scare

Melanoma cancer cases have increased more than 800 percent in the United States since statistics were first compiled in 1935. But what do CFCs and ozone depletion have to do with this? Absolutely nothing. Much of the apparent increase is due to better paperwork: Reporting of cases has improved over the years.

What other factors have caused the dramatic increase in the life-threatening form of skin cancer?

First, some facts about skin cancer. Skin cancer is a disease that afflicts lighter-skinned individuals. Table 6.1 shows that darker-

Table 6.1
CLASSIFICATION OF SUN-REACTIVE SKIN TYPES

Skin color	Skin type	History of sunburning or tanning	Cancer risk
	I	Always burns easily, never tans	High risk
	II	Always burns easily, tans minimally	High risk
White	III	Burns moderately, tans gradually and uniformly	Moderate risk
	IV	Burns minimally, always tans well	Low risk
Brown	V	Rarely burns, always tans well	Very low risk
Black	VI	Never burns, deeply pigmented	Extremely low risk

skinned individuals (Orientals, Indians, Hispanics, southern Europeans, and Blacks) have very little risk of skin cancer. In fact, skin cancers are so uncommon in darker-skinned people that incidence data are difficult to obtain.

Skin cancer rates have been rising among lighter-skinned populations of the Western countries, where lifestyles have changed dramatically over the past two generations. Sixty years ago, the seasonal change from winter to summer sunlight occurred gradually. Today, however, one can board a plane in New York City in December and be frying like a lobster on Miami Beach a few hours later. White skin is not designed to withstand such dramatic changes in sunlight exposure.

How serious a health threat does skin cancer represent? The most common form of skin cancer, basal cell carcinoma, is disfiguring if not treated, but rarely kills its victims. Malignant melanoma, the least frequently found skin cancer, has the highest rate of mortality. Malignant melanoma accounts for only 4 percent of all skin cancer cases, but 75 percent of skin cancer deaths.

Malignant melanoma is also the least frequent of all the skin cancer types associated with overexposure to sunlight. A study by the Council on Scientific Affairs of the American Medical Association, published in the *Journal of the American Medical Association* July 21, 1989, summarizes scientific findings on this point:

> Despite the positive correlations relating the incidence of cutaneous malignant melanoma to [ultraviolet radiation] exposure, it is obvious that factors other than [ultraviolet radiation] are involved. Unlike non-melanoma skin cancer, which has a greater incidence in older individuals, cutaneous malig-

nant melanoma is most common during the middle decades of life. Non-melanoma skin cancer occurs most often in outdoor workers, whereas cutaneous malignant melanoma affects a greater relative percentage of urbanites who work indoors. The incidence of cutaneous malignant melanoma does not correlate well with latitudinal gradients [that is, sunshine] in Western Australia and Central Europe. Anatomic distribution of cutaneous malignant melanoma does not *closely* match body areas of greatest Sun exposure, as it does for non-melanoma skin cancers. Histologically, relatively little solar elastosis occurs in the vicinity of cutaneous malignant melanoma, whereas it is closely associated with basal cell carcinoma.... No animal model has been developed that allows cutaneous malignant melanoma to be consistently induced by ultraviolet radiation alone [p. 382].

The report's bottom line is that medical science has no idea of what causes malignant melanoma.

One new hypothesis, put forward by Cedric Garland and Frank Garland from the University of California's Cancer Epidemiology Department in San Diego, is that the widespread and growing use of sunscreens over the past two decades is the cause of the rise in skin cancer. According to Frank Garland, it may be the near-ultraviolet (ultraviolet A, 320 to 400 nm) that is causing the melanomas, even though middle range (ultraviolet B, 286 to 320 nm) has always been considered the harmful range of the ultraviolet spectrum. This is because malignant melanoma arises in the melanocyte cells of the skin's dermis layer—and less than 10 percent of ultraviolet B reaches the dermis, while more than 50 percent of the ultraviolet A radiation does. Moreover, modern sunscreens are designed to block ultraviolet B, yet are transparent to the supposedly benign ultraviolet A.

But doesn't overexposure to ultraviolet B harm the skin by causing sunburn? Yes, it does. Sunburn is the body's alert mechanism to warn the individual that he has spent too much time in the Sun. Individuals who wear sunscreen short-circuit that alert mechanism and subject their bodies to massive doses of ultraviolet A radiation; in the case of a vacationer on a southern beach, this can be perhaps as much as 50 times the sunlight his body is accustomed to. Furthermore, sun tanning, which is the long-term protective mechanism of the body against excessive sunlight exposure, is also triggered by ultraviolet B. By selectively filtering out ultraviolet B, suncreens are foiling not only the body's alert mechanism but also the long-term protection provided by suntan and a thicker dermis.

Another danger of ultraviolet A comes from tanning parlors, which produce a suntan by irradiating customers with the allegedly benign ultraviolet A radiation. The problem is that by eliminating the allegedly harmful ultraviolet B, again the body's alarm mechanism is foiled, and the customers are irradiated with 100,000 to 300,000 times the concentration of ultraviolet A that they would receive if they stayed out in the Sun for the same amount of time. Since 1 million Americans use tanning parlors every day, this is a significant problem.

Sunscreens also screen out some of the most beneficial wavelengths in the ultraviolet B spectrum, including those that synthesize Vitamin D3 hormone and trigger the repair function of the body, which fights infections, repairs damaged cells and DNA, and even fights some kinds of cancers.

How do the dangers of skin cancer stack up against the dangers from other kinds of cancer? As a percentage of the 510,000 annual deaths from cancer, the 8,800 deaths from skin cancer represent barely 1.7 percent of the total. Of course, it is terrible that even one individual dies of skin cancer. The point is that there are significantly more serious health hazards than skin cancer. Trillions of dollars are being spent to deal with this alleged skin cancer threat while comparatively little funding is going toward basic research and programs to combat the more serious threat of cancer in general, and infectious diseases such as AIDS, tuberculosis, and cholera.

The U.S. government's Centers for Disease Control in Atlanta released a report Jan. 31, 1990, announcing that more than 434,000 Americans had died in 1988 from health problems caused by smoking, an 11 percent increase over 1985. Smoking is the number one cause of preventable deaths in the United States, responsible for about one-fifth of all deaths. The estimates include death from various cancers, heart disease, high blood pressure, stroke and lung diseases, as well as burn deaths and infant deaths caused by conditions related to a mother's smoking during pregnancy.

In other words, smoking kills 50 times more Americans than skin cancer, yet one rarely hears an outcry about cigarette smoking from the environmental movement. But these same environmentalists do not hesitate to demand hundreds of billions of dollars to cut smoke emissions from factories, although in the United States, more than 92 percent of all pollutants inhaled by people (smoke, carbon monoxide, and so on), come from cigarette smoking, while the amount of pollutants inhaled by Americans from factory emissions is almost nil. Is it just a coincidence that the tobacco industry, including the Reynolds family, is one of the largest funders of the environmental movement?

Are the environmentalists really concerned about the state of public health in America? According to a report released in February 1991 by the General Accounting Office, more than 100 U.S. hospitals are closing every year. Yet, President George Bush, with enthusiastic support from the environmentalist movement, has made climate change and so-called clean air the number one priorities of his domestic agenda. Shortly after President Bush signed the 1990 Amendments to the 1970 Clean Air Act, the Environmental Protection Agency released its much delayed report on the costs of environmental legislation. The report revealed that environmental laws on the books as of 1989 were costing the U.S. economy more than $115 billion a year. The total costs since 1972 were calculated at more than $1.2 trillion dollars—and even this figure is probably vastly understated. The latest official numbers indicate that environmental costs have risen to $131 billion a year, for a total investment of $1.6 trillion since 1972.

A more realistic estimate, however, is that the cost of environmental legislation was more than $200 billion in 1990. This cost will rise significantly as a result of the signing of the Clean Air Act, the ban on CFCs, and the hundreds of other environmental laws passed during 1990. This money would be much better spent on rebuilding America's collapsed health care system.

The Dosage Question

Dosage is a fundamental, and nonlinear, phenomenon in biology. Many things we axiomatically regard as natural and beneficial for one's health, such as vitamins, are dosage-dependent. Vitamin A, for example, is necessary to the body, but too much of it leads to nerve, skin, liver, and bone disorders.

One unfortunate result of environmental propaganda over the past two decades has been the growing hegemony of the belief that if a substance is toxic in high concentrations, it will be toxic in very low concentrations as well. This way of thinking led millions of mothers to panic over the presence of minuscule amounts of the chemical Alar in the apples they were feeding their children.

Unfortunately, most people are unable (or refuse) to think in terms of proper scientific method. The fact that a substance is toxic in high concentrations does not mean it will be toxic in low concentrations. For example, the science of vaccination is based on giving people a small dosage of a virus or bacteria. This triggers an immune response from the body, which generates immunity to that organism. Because of the low concentrations, the body can easily overwhelm the intruder. Therefore, when a full-blown infection strikes, the body is prepared to defend itself. A better example is aspirin: Two will relieve a headache. But 2,000 at one sitting will kill you, while 2,000 taken one a day for five and one-half years may help prevent colon and breast cancer, according to the latest research.

This point is entirely missed in the controversy over ultraviolet radiation. There is no question that massive overexposures to both ultraviolet A and ultraviolet B can severely damage the human body. But the skin color of all human beings has adjusted to relatively small amounts of sunlight found at certain geographical latitudes; as long as that natural ratio is maintained, there is little danger of damage. Furthermore, ultraviolet radiation is *necessary* for the body's metabolism. Lack of ultraviolet radiation may be more dangerous than too much of it.

Ultraviolet light shorter than 290 nm is toxic to all forms of *unpigmented* living cells. How then do living things on the Earth survive exposure to incoming ultraviolet radiation? Plants and algae have a variety of photosynthetic pigments: chlorophylls, caratenoids, phycobilins, and fucoxanthin; these absorb electromagnetic radiation even into the ultraviolet range and harness its energy in ways useful to life. Most animals have fur, feathers, or scales that offer protection. Fish are partially shielded from ultraviolet radiation by water. And man has a skin pigment: melanin.

The same ultraviolet radiation that is deadly in large doses to unprotected living cells, however, is necessary to all life on this planet. No living thing—not plants, not animals, not man—could long survive if ultraviolet radiation were totally screened out of the Earth's atmosphere.

In human beings, normal exposure to ultraviolet light triggers the conversion of one of the cholesterols (7-dehydrocholesterol) found in the skin to calciferol, or vitamin D, a necessary nutrient. Deficiency in vitamin D leads to rickets and disturbances in calcium and phosphorus metabolism. On the basis of studies of the X-rays of fossilized bones of the Neanderthal man, scientists have now hypothesized that many of the features that made Neanderthal man appear so different—the large head, the bulging forehead, the short hunched-over stature, and the frequent bowing of limbs— were caused by severe vitamin D deficiency: rickets. (See Curtis 1968.) This hypothesis is backed up by the observation that the more extreme fossil specimens, in terms of these types of deformities, come from geographic areas above the latitude of 40 degrees, an area in which sunlight is not abundant in the winter.

What dosage, or concentration, of ultraviolet light in the form of exposure to direct or indirect sunlight is beneficial to human beings? The dosage varies from individual to individual, depending on skin type. For the rare pigmentless albino, or for the relatively small percentage of Caucasians with exceedingly fair and poorly pigmented skin, prolonged exposure to direct sunlight may well be quite problematic. But for most Caucasians and, obviously, for the majority of humanity that are still more richly endowed with melanin, exposure to a mixture of direct and indirect sunlight in the course of outdoor activity is a health-promoting experience, as long as the length of exposure is kept below the level at which sunburn and sun-poisoning occur.

How large a dosage of ultraviolet light an individual can tolerate without sunburn or sun-poisoning is also dependent upon the tan acquired during previous exposure. The faddish ritual of all-day exposure to direct sunlight on weekends in early summer, and the subsequent sunburn and possible sun-poisoning that follows, is dangerous. Because the population tends to be irrational, following a fad of Sun worship rather than health interests, many medical professionals tend to err on the side of stressing the dangers of overexposure to the ultraviolet wavelengths in sunlight, rather than the benefits of low dosages of exposure.

In addition to the cancer scare, the ozone depletion theorists insist that the increase in ultraviolet light reaching the Earth's surface because of ozone depletion is causing an increase in the incidence of cataracts. On this point, the environmentalists have no scientific evidence. For one, there is no epidemiological evidence that shows an increase in the rate of cataracts as one moves closer to the equator, with a concomitant increase in ultraviolet B radiation.

It is important to examine the experimental methodology that led to the cataract scare. The claims of cataract damage from ultraviolet B resulted from a series of experiments in which animals were subjected to very intense amounts of ultraviolet B radiation aimed directly at their eyes. Again we return to our magnifying glass example. By focusing the Sun's rays on the skin, one would cause a very severe burn on the skin in a few minutes. However, this does not mean that standing out in the Sun is dangerous.

The Benefits of Ultraviolet Light

One of the major functions of ultraviolet radiation is as nature's cleaner. Together with low-level ozone, ultraviolet light is the most effective germicidal and bactericidal agent in the atmosphere. Because ultraviolet light is very toxic to all forms of unpigmented living cells, particularly unicellular plants and animals, it kills germs, viruses, bacteria, and other microorganisms. Without it, there would be no natural check on the spread of deadly pandemics.

In 1877, two medical doctors, Downes and Blunt, accidentally made the discovery that ultraviolet light kills bacteria. Observing uncolored tubes of brown sugar water, which they had placed on a window sill, they found that the tubes in the shade had become cloudy, indicating bacterial growth. The tubes exposed to the light had remained clear, indicating no bacterial growth. "The most marked differences in the two sets of tubes were obtained when the Sun shone brightly. Light," they decided, "is inimical to the development of bacteria."

Fifteen years later, in 1892, scientist Marshall Ward demonstrated that the ultraviolet portion of the electromagnetic spectrum has the most intense antibacterial properties. Following Ward's lead, other scientists discovered the ability of ultraviolet light to kill many specific kinds of bacteria. Table 6.2 presents the dates of discovery of ultraviolet sensitivity of different bacteria.

Human beings and lower animals all benefit from moderate doses of ultraviolet light. Most vertebrates require vitamin D for proper development of the skeletal system. For many organisms, the only source of this vitamin is through the action of ultraviolet light on calciferol in their outer covering of skin, fur, or feathers—

Table 6.2
DISCOVERY OF SUN-SUSCEPTIBILITY OF BACTERIA

Bacterium's common name	Bacterium's scientific name	Date	Scientist
Anthrax	*Bacillus anthracis*	1886	Arloing
Plague coccobacillus	*Pasteurella pestis*	1887	Polerino
Strep	*Streptococcus*	1887	Duclaux
Tubercle bacillus	*Mycobacterium tuberculosis*	1890	Koch
Cholera	*Vibrio comma*	1892	Moment
Staph	*Staphylococcus*	1892	Chemelewsky
Colon bacillus	*Escherichia coli*	1894	Dieudonne
Dysentery bacillus	*Shigella dysenteriae*	1909	Henri

a route not open to fish, who thus had to evolve a way of manufacturing vitamin D without sunlight.

Scientists have long known that moderate exposure to sunlight can cure rickets, a bone disease caused by insufficient creation of vitamin D in the body. W. F. Loomis, writes in a 1970 article in *Scientific American:*

> The discovery of the cause and cure of rickets is one of the great triumphs of biochemical medicine, and yet its history is little known. Indeed, it is so little known that even today most textbooks list rickets as a dietary-deficiency disease resulting from a lack of "vitamin D." In actual fact rickets was the earliest air-pollution disease. It was first described in England in about 1650, at the time of the introduction of soft coal, and it spread through Europe with the Industrial Revolution's pall of coal smoke and the increasing concentration of poor people in the narrow, sunless alleys of factory towns and big-city slums. This, we know now, was because rickets is caused not by a poor diet but by a deficiency of solar ultraviolet radiation, which is necessary for the synthesis of calciferol, the calcifying hormone released into the bloodstream by the skin. Without calciferol, not enough calcium is laid down in growing bones, and the crippling deformities of rickets are the consequence. Either adequate sunlight or the ingestion of minute amounts of calciferol, or one of its analogues, therefore prevents and cures rickets, and so the disease has been eradicated [p. 77].

Not only humans are affected by rickets. Animals held in captivity in cages without a lot of sunshine or in polluted cities also develop rickets. Loomis reports: "As early as 1888 the English physician Sir John Bland-Sutton found unmistakable evidence of rickets in animals in the London zoo—chimpanzees, lions, tigers, bears, deer, rabbits, lizards, ostriches, pigeons and many other species. Bland-Sutton noted that 'in spite of every care and keeping them in comfortable dens' lions in London developed rickets, whereas 'in Dublin, Manchester, and some other British towns, lions can be reared successfully in captivity.' It is clear in retrospect that the pall of coal smoke over London was the causative factor" (p. 77).

Loomis also reports the fascinating results of the work of English medical missionary Theobald Palm during an 1890 trip to Japan. Palm noted the virtual absence of the disease among the Japanese, and was spurred to begin collection of a data base on the international prevalence of rickets. Palm's data soon showed that rickets "was essentially confined to northern Europe and was almost completely absent from the rest of the world," (p.79). Palm concluded that rickets was caused by the absence of sunlight: "It is in the narrow alleys, the haunts and playgrounds of the poor, that this exclusion of sunlight is at its worst, and it is there that the victims of rickets are to be found in abundance." Palm recommended "the systematic use of sun-baths as a preventive and therapeutic measure in rickets" [p. 79].

Although the link between rickets and lack of sunlight was established before the turn of the century, it was not until 1919 that it was discovered that the benefits of sunlight in preventing and treating rickets came from the ultraviolet portion of the electromagnetic spectrum. This discovery was made by Berlin pediatrician Kurt Huldschinsky. His work also put medical science on the track of the hormone involved in vitamin D synthesis, calciferol. Loomis concludes:

> The recognition of calciferol as an ultraviolet-dependent hormone gives fresh meaning to a number of seemingly unrelated physiological and cultural adaptations. Tropical man probably avoids the dangers of too much calciferol production by virtue of his dark skin; the melanin granules in the outer layers protect the lower layers of the skin. European man, on the other hand, needed to use all the scanty ultraviolet light available, and consequently was gradually selected for an unpigmented skin such as is present in extreme degree in the blond-haired, blue-eyed, fair-skinned and rosy-cheeked

175

infants of the English, north German and Scandinavian peoples. . . .

June weddings tend to bring the first baby in the spring; an infant born in the fall was almost certain to have rickets by the time he was six months old. The fish-on–Friday tradition was as adaptive as the scurvy-preventing eating of an apple a day. Taking the baby out of doors even in the middle of winter for "some fresh air and sunshine" became a northern folk custom [pp. 90–91].

Atmospheric scientist Hugh Ellsaesser maintains that "Rickets is but one of the hazards of insufficient ultraviolet." Writing in the Summer 1990 *21st Century Science and Technology* magazine, Ellsaesser pinpoints lack of sufficient ultraviolet radiation as the cause of osteomalacia, a disease of aging:

At present the most serious health hazard in the United States from insufficient ultraviolet or "vitamin D" is osteomalacia, or wasting bone loss in the elderly. While this process can be arrested or slowed by proper treatment, the best treatment appears to be to assure that there is both adequate "vitamin D" and minerals available during the growth period while the skeleton is forming. . . .

Bone fracture, particularly of the femur, among the elderly suffering from osteomalacia is a far more serious health problem than ordinary skin cancer. There are some 400,000 to 600,000 new cases of skin cancer per year in the United States, while among the 20 million Americans affected by osteomalacia there are more than 1,200,000 bone fractures each year.

These statistics strongly suggest that any increase in ultraviolet resulting from ozone loss would, at least eventually, exert a beneficial impact on our health greater than the detrimental one now emphasized. This becomes even more credible when it is recalled that our bodies are far more capable of letting us know when we are getting too much ultraviolet than they are at letting us know when we are getting too little [p. 9].

Doctors Frank and Cedric Garland and Edward Gorham have come up with some interesting hypotheses concerning the benefits of ultraviolet exposure, resulting from their studies of the distribution of occurrences of colon and breast cancer. They have found inverse correlations between mortality rates from colon and breast cancer and the amount of both vitamin D and total sunlight. That is, the more sunlight available, including ultraviolet and vitamin D, the lower the mortality rates from these forms of cancer. It is very likely that other diseases could be found that show a negative correlation with exposure to ultraviolet light. Perhaps more progress could be made in finding a cure for cancer if the government were willing to spend as much money on biomedical research as it has spent investigating the ozone hole and alleged threats to the ozone layer.

A huge clinical trial being planned by the National Institutes of Health (NIH) may prove the point made by the Drs. Garland and Gorham. According to Geoffrey Cowley, writing in *Newsweek,* Dec. 30, 1991, approximately 60,000 postmenopausal women participating in an NIH study to begin this year will receive either a calcium and vitamin D supplement twice a day, or a placebo for up to nine years. Dr. William Harlan, associate director for disease prevention at NIH, reports that the goal of the project is to reduce colon cancer and bone fractures by 25 percent. The participants, however, will also be monitored for a reduction in breast cancer and other conditions. If the Garlands and Gorham are correct, vitamin D (that is, the hormone produced by ultraviolet radiation), will reduce the incidence of both colon and breast cancer.

We have now entered one of the most fascinating realms of biology and medicine today: the effect of light on the human body, known as photobiology or optical biophysics. This includes the health benefits of full spectrum light (as opposed to present-day fluorescent lighting, which emits a highly truncated and possibly harmful spectrum of light). The most promising medical breakthroughs in curing diseases like cancer and AIDS lie in this field of scientific endeavor. Although space here does not permit more than a glimpse of the very exciting research that exists today, we have included an expanded reference list for those who wish to pursue the topic further. (See especially Kime and Liberman.)

References

Robert Day Allen, 1977. *The Science of Life.* New York: Harper and Row, Publishers, Inc.

Valerie Beral, Helen Shaw, Susan Evans, and Gerald Milton, 1982. "Malignant Melanoma and Exposure to Fluorescent Lighting at Work," *Lancet,* (Aug. 7), pp. 290–93.

Council on Scientific Affairs, American Medical Association, 1989. "Harmful Effects of Ultraviolet Radiation," *Journal of the American Medical Association,* Vol. 262, No. 3, (July 21): pp. 380–384.

Geoffrey Cowley, 1991. "Can Sunshine Save Your Life? Vitamin D May Help Fight Colon and Breast Cancer." *Newsweek,* (Dec. 30), p. 56.

Helena Curtis, 1968. *Biology.* New York: Worth Publishing Inc.

Arne Dahlback, Thormod Henriksen, Søren H.H. Larsen, et al., 1989. "Biological UV-Doses and the Effect of an Ozone Layer Depletion," *Photochemistry and Photobiology,* Vol. 49, No. 5 (May): pp. 621–625.

Cedric F. Garland, 1990. "Acid Haze Air Pollution and Breast and Colon Cancer Mortality in 20 Canadian Cities," *Canadian Journal of Public Health,* Vol. 80 (March–April), pp. 96–100.

Cedric F. Garland and Frank C. Garland, 1980. "Do Sunlight and Vitamin D Reduce the Likelihood of Colon Cancer?" *International Journal of Epidemiology,* Vol. 9, pp. 227–231.

————, 1986. "Calcium and Colon Cancer," *Clinical Nutrition,* Vol. 5, (July/August): pp. 161–166.

Cedric Garland, Richard B. Shekelle, Elisabeth Barrett-Connor, et al., 1985. "Dietary Vitamin D and Calcium and Risk of Colorectal Cancer: A 19-year Prospective Study in Men," *The Lancet* (Feb. 9), pp. 307–309.

Cedric Garland, George Comstock, Frank Garland, et al., 1989. "Serum 25-Hydroxyvitamin D and Colon Cancer: Eight-Year Prospective Study, *The Lancet* (Nov. 18), pp. 1176–1178.

Frank Garland, Cedric Garland, and Jeffrey Young, 1990. "Geographical Variation in Breast Cancer Mortality Rates and Solar Radiation in the United States," *Preventive Medicine* (December), pp. 614–622.

Frank Garland, Cedric Garland, and Edward Gorham, 1990. "Sunlight, Vitamin D, and Breast Cancer Incidence in the U.S.S.R.," *International Journal of Epidemiology* (December).

Fritz Hollwich, 1979. *The Influence of Ocular Light Perception on Metabolism in Man and in Animal.* New York: Springer-Verlag.

Zane R. Kime, 1980. *Sunlight Could Save Your Life.* Penryn, Calif.: World Health Publications.

Jacob Liberman, 1990. *Light, Medicine of the Future.* Santa Fe, N.M.: Bear & Company.

Michael J. Lillyquist, 1985. *Sunlight and Health.* New York: Dodd, Mead & Company.

W.F. Loomis, 1970. "Rickets," *Scientific American* (December), p. 76–86.

John Ott, 1973. *Health and Light.* Greenwich, Conn.: Devin-Adair.

————, 1990. *Light, Radiation and You.* Greenwich, Conn.: Devin-Adair.

Stuart A. Penkett, 1989. "Ultraviolet levels down not up," *Nature,* Vol. 341 (Sept. 28), pp. 283–284.

Fritz-Albert Popp, ed., 1989. *Electromagnetic Bio-Information*. Baltimore: Urban & Schwarzenberg.

Joseph Scotto, Gerald Cotton, Frederick Urback, et al., 1988. "Biologically Effective Ultraviolet Radiation: Surface Measurements in the United States, 1974–1985," *Science,* Vol. 239 (Feb. 12), pp. 762–764.

Joseph Scotto, 1988. "Global Stratospheric Ozone and UVB Radiation," *Science,* Vol. 239 (Nov. 25), pp. 1111–1112.

7

The Montreal Protocol

From June 27 to 29, 1990, the representatives of 93 nations met in London at a conference chaired by Margaret Thatcher, then British prime minister. The meeting was called to revise the original Montreal Protocol of September 1987, which had set the first global controls on the manufacture of chlorofluorocarbons. Of the 93 nations present, 59 agreed to sign a new version of the Montreal Protocol, which imposed even more draconian cutbacks on CFCs than the original treaty and added several more chemicals to the list of chemicals to be banned. Some 34 nations, however, refused to sign the protocol, arguing that there was not enough evidence to indict these compounds.

By summer 1991, more than a year later, the euphoria that accompanied the signing of the treaty had died down, replaced by the sobering realization that it will be much more difficult and costly to replace CFCs than originally thought. The American Society of Heating, Refrigeration, and Air Conditioning Engineers (ASHRAE) warned in a special issue of its journal, February 1991, that the prospects for CFCs replacements were dismal. This is a sobering statement coming from a publication that has been one of the staunchest supporters of the ban on CFCs.

Indeed, there is now a realization that the cost of banning CFCs and other halogenated chemicals may be overwhelming to the world economy. Estimates from experts in different industries that will be affected by the ban indicate that the cost may be as high as $5 trillion by the year 2005.

More significant than the cost in dollars and cents of banning CFCs is the cost in human lives. The increase in human population in the 20th century is primarily the result of public health policies and the improved availability of food, which has come about in large part because of the extraordinary quality of CFCs as refrigerants. More than 75 percent of the food consumed by Americans today is refrigerated at some point by CFCs. The ban on CFCs will mean that most of the hundreds of millions of refrigeration units installed worldwide will have to be scrapped. The consequences of this will be a collapse of food storage capacity worldwide and a dramatic increase in the death rate from hunger, starvation, and food-borne diseases. Experts on the worldwide food chain estimate that between 20 and 40 million individuals will die *every year* as a result of the ban on CFCs. The only feasible alternative to CFCs for efficient food preservation is sterilization by food irradiation, which is under attack by the same environmentalists who are out to ban CFCs.

News articles reporting on the London Conference of June 1990 claimed that the representatives of the 59 nations that ratified the treaty signed on because of the compelling scientific evidence presented at the conference. This could not be further from the truth. Most nations that signed did so because they had a gun to their heads. There is a clause in the Montreal Protocol that mandates economic warfare against any nation that does not participate. Article 4 of the treaty, titled "Control of Trade with Non-Parties," describes in detail the actions that will be taken against nations that do not sign the protocol. The clause mandates that signatory nations will impose a total trade embargo against any nation of the world that does not abide by the Montreal Protocol. Signatory nations are even prohibited from exporting these "controlled substances" to any nation that does not sign the protocol. Faced with such a murderous clause, most nations have no alternative but to sign the treaty.

Why is such a clause necessary in a treaty that is supposed to save the Earth, and thus save human lives? The reason is simple: The banning of CFCs will have a devastating impact on modern industrial society. Those nations that are trying to industrialize

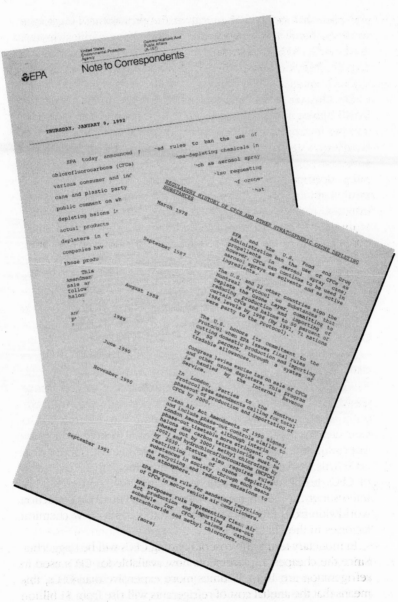

EPA
United States
Environmental Protection
Agency
Communications And
Public Affairs
(A-107)

Note to Correspondents

THURSDAY, JANUARY 9, 1992

EPA today announced [   ] -ed rule to ban the use of
chlorofluorocarbons (CFCs) [   ] -ne-depleting chemicals in
various consumer and ind[   ] -ch as aerosol spray
cans and plastic party [   ] -lso requesting
public comment on wh[   ] -f ozone-
depleting halons ir [   ] -hat
actual products

REGULATORY HISTORY OF CFCS AND OTHER STRATOSPHERIC OZONE DEPLETING SUBSTANCES

depleters in t[   ]
companies hav[   ]
those produ[   ]

This [   ]
Amendment [   ]
sale ar [   ]
follow [   ]
halon[   ]

and [   ]
p[   ]
r[   ]

March 1978

EPA and the U.S. Food and Drug
Administration ban the use of CFCs as
propellants in aerosol spray cans;
however, CFCs can continue to be used in
aerosol sprays as solvents and as active
ingredients.

September 1987

The U.S. and 22 other countries sign the
Montreal Protocol on Substances that
Deplete the Ozone Layer, committing to
reducing the production and importing of
certain CFCs and halons to 50 percent of
1986 levels by 1998 (by 1991, 71 nations
were party to the Protocol).

August 1988

The U.S. honors its commitment to the
Protocol when EPA issues final rules
cutting domestic production and importing
by 50 percent, through a system of
tradable allowances.

1989

Congress levies excise tax on sale of CFCs
and other ozone depleters. This program
is handled by the Internal Revenue
Service.

June 1990

In London, Parties to the Montreal
Protocol pass amendments calling for total
phaseout of production and importation of
CFCs by 2000.

November 1990

Clean Air Act Amendments of 1990 signed,
and include phase-out controls similar to
London Amendments, although more stringent.
phase-out timetable for total phase-out of
halons and carbon tetrachloride. CFCs,
phased out by 2000; tetrachloride must be
by 2002 and by hydrochlorofluorocarbons by
2030. Statute also requires
restricting use of substances in society, through depleting
as recycling and reducing emissions to
the atmosphere. rules
ozone depleting such means

September 1991

EPA proposes rule for mandatory recycling
of CFCs in motor vehicle air conditioners.

EPA proposes rule implementing Clean Air
Act production and importing phase-out
schedule for CFCs, halons, carbon
tetrachloride and methyl chloroform.

(more)

The EPA phased out the use of CFCs without taking into account the
consequences in terms of human suffering.

182

are essentially being told to stay in the preindustrial dark ages. Similarly, families in the lower and middle classes of the industrialized nations will pay a heavy penalty, sacrificing their standard of living on behalf of an emerging international cartel that will control the technologies of the future.

The environmental hoaxsters behind the ban on CFCs claim that it will be simple and easy to replace these chemicals. This is a lie. The public has been told that there are "ozone-friendly" chemicals that can be "dropped into" currently existing equipment to replace CFCs. No one has yet demonstrated this to be true. There are no such "drop-in" replacements. This means that all of the existing equipment that uses CFCs will have to be junked and replaced by equipment that can use new chemicals. Given the fact that the equipment to be scrapped includes hundreds of millions of home, commercial, and industrial refrigerators, it is not a small matter.

The equipment to be scrapped includes:
• 610 million refrigerators and freezers
• 120 million cold storage units
• 100 million refrigerated transports
• 150 million car air conditioners.

The infrastructure problem does not end here. Extremely important is the issue of those refrigerators that will not be built, or will be built at a much higher cost. Countries in areas of the world where fewer than 1 in 100 households has a refrigerator—India, China, Brazil, South Korea, and Taiwan, for example—had embarked on ambitious programs to produce refrigerators. Under estimates made in 1988, these countries were expected to purchase or produce between 400 and 500 million refrigerators by the year 2000. This would have required a sevenfold increase in the amount of CFCs produced every year for refrigeration purposes. Most interestingly, it would have shifted the bulk of CFC production worldwide from Europe and the United States to new chemical factories in the Third World.

In monetary terms, the cost of banning CFCs will be staggering. Since the cheapest replacements now available for CFCs used in refrigeration are 10 to 30 times more expensive than CFCs, this means that the annual cost of refrigerants will rise from $1 billion per year in 1988, to between $10 and $30 billion per year, if all CFCs used in refrigeration are replaced. With a fivefold increase in refrigerants by the year 2000, the tab will be between $50 and $150 billion per year. This figure is consistent with estimates made by the refrigeration industry that the annual market in refrigerants will be between $150 and $200 billion a year by the year 2000.*

Refrigeration industry experts have privately acknowledged that the cost of banning CFCs will be in the range of $500 billion to $1 trillion by the year 2000—for the refrigeration industry alone. This figure does not include the increased cost of foods caused by increased refrigeration costs. Who will pay the tab? Who will pocket the profits?

Moreover, replacements for CFCs are not easy to find. The main replacements were expected to come from HCFCs, chemicals closely related to CFCs. The difference between the two types of compounds, which allegedly makes HCFCs "environmentally benign," is the presence of an additional hydrogen atom. HCFCs still have that allegedly ozone-busting chlorine atom, but the extra hydrogen atom makes HCFCs less stable than their CFC cousins and therefore they supposedly break down before reaching the ozone layer. In techno-lingo, this means that HCFCs have a lower ozone depletion potential or ODP than CFCs have. (ODP is one of the most absurd concepts promulgated by the ozone depletion theorists. Because no one has ever actually observed a CFC or a halon break down in the stratosphere, how can one actually measure such a hypothetical scenario?)

Since the signing of the Montreal Protocol, HCFCs have come into disfavor. During several meetings of the Intergovernmental Panel on Climate Change, HCFCs were attacked because they are supposedly "super-greenhouse" gases. The panel discussed how HCFCs should be banned by the year 2000, instead of the target dates of between 2020 and 2040 set by the Montreal Protocol and the Clean Air Act signed by President George Bush in 1990.

More recently, the Institute for Energy and Environmental Research, an environmentalist think thank in Washington, released a study claiming that HCFCs are three to five times more ozone-destructive than previously estimated. This announcement, reported in *The Washington Post* Feb. 23, helped set the stage for a huge environmentalist onslaught in Washington (see below). Senator Albert Gore demanded that HCFCs be banned by the year 1999, arguing, according to *The Washington Post:* "It's a mistake

---

* We are comparing the cost of replacements to the cost for CFCs in 1988. The reason is that the CFC tax and increasing shortages have driven up the price of CFCs so much that replacements are now only 10 times more expensive and in some cases have the same price as CFCs.

to replace one ozone-destroying chemical with another. Any added increments at this point represent true insanity" (p. A3).

"True insanity"? For industrial and domestic users, the impact of an early ban on HCFCs will be profound. Just in the United States, HCFCs are needed to service an estimated $200 billion worth of equipment ranging from home air conditioners and heat pumps to cold storage cases at supermarkets. Furthermore, industry was counting on switching from CFCs to HCFCs in many uses as a transition until other replacements are found. Many industries acquiesed to the ban on CFCs because they were promised that HCFCs would be available to replace them.

Murder of the Third World

One of the immediate results of banning CFCs will be to drive Third World chemical producers out of business. Third World chemical industries do not have the research capabilities to manufacture alternatives to CFCs, and those alternatives in existence will be patented to members of the chemical cartel in the West. Even if the major companies sell their patents to developing countries (at hefty prices), these countries will still have to scrap their chemical plants and build new ones designed to produce these new chemicals.

This is a question of what has been called technological apartheid. In March 1989, spokesmen for several large chemical corporations announced that construction of chemical factories for production of CFCs in the Third World had been halted, and existing contracts to build CFC factories would not be honored! The new policy is to export CFCs to these nations for a few years until "phaseout," when Third World countries will be allowed to purchase the expensive substitutes—and the new equipment needed to use them.

This policy means that millions will die.

Consider what will happen if all the refrigeration systems of the world shut down. This can be imagined by examining how Americans lived before the advent of refrigeration. In her excellent book, *Trashing the Planet,* former Washington governor and Atomic Energy Commisssion head Dixy Lee Ray gives a vivid account of life without refrigerators 60 years ago:

> When I was young, fresh food could be eaten only when in season. Some fruits and vegetables could be preserved by

H. Dalrymp/UNICEF

A Ugandan family overcome by starvation. How many people will die across the developing sector, as the result of the ban of modern agricultural, industrial, and medical chemicals?

canning, but the processes for safe home canning were poorly understood and unreliable. Food poisoning was common. Every year people were stricken with botulism and salmonella or various digestive upsets caused by eating decayed or tainted food. During the winter, therefore, only potatoes, carrots, cabbage, winter squash, onions, and dried beans were generally available. Some people ate rutabagas, turnips, and parsnips, but, in our family, these were considered "horse food." Only with the introduction of widespread refrigeration of rail cars and boats, warehouses, and grocery food containers, could fresh food be brought to the customer throughout the year. Refrigeration vastly expanded the market for all foods—meat, dairy products, fruit and vegetables... [p. 16].

Today, refrigeration, like food, is taken for granted in the advanced-sector countries. Most people do not realize that in the rest of the world, a large proportion of the food produced spoils before it can be consumed due to lack of refrigeration. Many of the 40 million people who starve to death every year did not die because there was a lack of food. They starved because the food supplies abundant at the time of the year that the crops were harvested had spoiled before they were needed.

One example of the problem with food spoilage today is the former Soviet Union, where transportation bottlenecks prevent food from moving efficiently around the country, and a large percentage of what is produced, in some cases as much as 65 percent, spoils in transit or before transit.

The importance of adequate refrigeration for increasing the world food supply was discussed by Professor W. Kaminski of the Institute of Agricultural and Foodstuff Economy in Warsaw, Poland, at a 1988 international refrigeration conference in Paris. Kaminski reports: "Refrigeration—under condition that a well organized and complex cold chain is applied—can, to a large extent, contribute to increasing the safeguarding of world food resources through ensuring an important decrease in the quantitative and qualitative losses in food produced, right from the harvesting of raw materials up to the consumption of the finished product."

The food losses Kaminski refers to are enormous:

It can be noted that the present world production of perishable products, and which necessitate refrigeration, is more than 1.5 milliard [1.5 billion] tonnes per year, of which 250 to 300 million tonnes are lost because refrigeration was not fully applied.... If we can safeguard these food products, there will be a supplementary food quantity of about 80 kg per inhabitant of this Earth per year.

This means that the world already produces enough food to feed every man, woman, and child on the face of the Earth. What causes famine and starvation is lack of transportation and refrigeration of food from where it is produced to the point of consumption.

A fascinating section of Kaminski's speech concerns the crucial role of fish as a protein supply for hundreds of millions in the Third World, and the role of refrigeration in safeguarding fresh fish supplies. In the decades since World War II, Kaminski says, fisheries resources have increased four times, now representing about 16 kg per person worldwide, nearly 80 million tons a year as of 1983. Such a vast expansion was made possible through "the widespread application of chilling, particularly freezing in fishing boats, generally high sea trawlers (which considerably widened fishing territorial waters) as well as through the extension of the cold chain for the requirements of inland fish economy (cold rooms, specialized refrigerated transport, wholesale fish dealers, etc.)."

Kaminski then documents that the ban on CFCs will be devastating to the world's cold storage and transport infrastructure, the fishing fleets, and the refrigerated sea transport fleet, which consists of some 10,000 ships with a tonnage exceeding 10 million cubic meters, excluding containers.

International refrigeration experts privately estimate that hundreds of millions will die over the next 15 years as the international "cold chain" collapses because of the ban on CFCs. Their gruesome figures run from 20 to 40 million *additional* deaths from starvation, per year, by the year 2005.

Even Robert Watson, head of the Ozone Trends Panel, admitted that banning CFCs means killing people—unnecessarily. In a 1989 interview with syndicated columnist Alston Chase, Watson confesses that "probably more people would die from food poisoning as a consequence of inadequate refrigeration than would die from depleting ozone." During a February 1990 meeting of the Intergovernmental Panel on Climate Change in Washington, D.C., one of the authors asked Watson how he could support a ban on CFCs if he knew so many would die as a result. Watson responded that he had changed his mind about the murderous consequences of the CFC ban. The top representative of the Du Pont Company, he said, had convinced him that their replacements for CFCs would do the job, and Watson now insisted not a single person would die because of lack of refrigeration. When asked how he could trust any figures from Du Pont, since they stood to make so much money on the ban, an irate Watson responded, "Of course they are going to make enormous profits, what else did you expect?"

Preventing the Third World from building a refrigeration capacity is one of the stated purposes of the Malthusian environmentalists now making policy in Washington. Environmental Protection Agency chief William Reilly made this clear in July 1989 when he stated, "The prospect of seeing countries move forward with major development plans involving, as we heard in China, a proposal for 300 million new refrigerators possibly based on CFCs, makes very clear that we must engage them in this process [the ban on CFCs]." Reilly and the leading environmentalists, however, know quite well the consequences of destroying the worldwide refrigeration chain.

Today, world production of CFCs is approximately 1.1 million tons a year. As noted in the introduction, CFCs are one of the most benign and versatile chemicals ever invented. They are highly stable, nonflammable, nonexplosive, nontoxic, odor-free, noncorrosive and cheap—qualities that make them extremely useful in industries and households. Therefore, a wide array of uses has been found for them: as refrigerants and coolants, as inflating agents in foams and insulation, and as ingredients in industrial solvents, cosmetics, household products, and cleaners. Halons, a related group of chemicals also banned under the Montreal Protocol, are the most effective fire-fighting chemicals known to man, have important military applications, and play an essential role in the protection of electronic and computer equipment. Methyl chloroform plays a critical role as a solvent in the electronics and metalworking industries. These chemicals, and the timetable on which they will be banned are listed in Table 7.1.

Hysteria Speeds the Phaseout Timetable

Throughout 1991 and early 1992, the ozone hysteria in the media increased to such a high pitch that the ozone-depletion lobby was able to secure a ban on CFCs much sooner than required by the Montreal Protocol. On Dec. 31, David Doniger of the Natural Resources Defense Council, Liz Cook of Friends of the Earth, and Michael Oppenheimer of the Environmental Defense Fund released a petition for acceleration of the CFC phaseout. This petition was presented to the International CFC and Halon Confer-

Table 7.1
TIMETABLE FOR CHEMICAL BANS UNDER THE MONTREAL PROTOCOL

Controlled substance	Uses	Controls under 1990 update to Montreal Protocol	Controls proposed by the Natural Resources Defense Council
CFCs	Refrigeration, air conditioning, rigid and flexible plastic foams, solvent for electronics industry, aerosols	Complete phaseout by 2000	Complete phaseout by 1995
Halons	Fire extinguishers for hospitals, ships, aircraft, and computer rooms	Complete phaseout by 2000	Complete phaseout by 1992
HCFCs	Replacement for CFCs in refrigeration, foam blowing, and aerosols	Phaseout by 2020–2040	Phaseout by 2000
Carbon tetrachloride	Chemical feedstock for CFCs, solvents, pharmaceuticals, pesticides and some paints	Phaseout by 2000	Phaseout by 1992
Methyl chloroform	Solvent for precision metalworking and electronics industries	Phaseout by 2005	Phaseout by 1992
Methyl bromide	Most effective agricultural fumigant		Phaseout by 1993

ence on Dec. 3–5, 1991 in Baltimore, with approval of the conference sponsors, the Alliance for a Responsible CFC Policy. (This group, largely controlled by Du Pont, ICI, Allied Signal, and other chemical giants, was created in 1980 allegedly to defend CFCs. Under the leadership of Kevin Fay, however, it has become one of the most diligent proponents of a CFC ban, so long as such a ban occurs within a timetable that benefits the producers of alternative chemicals.)

On Feb. 3, 1992, NASA officials gave an emergency press conference warning of a potential ozone hole over North America because increased chlorine monoxide had been measured over the Northeast, according to data taken from one flyby mission that had not yet been fully analyzed. Within days of this headline-grabbing announcement, Senator Albert Gore, environmentalist groups, and the chemical industry had followed up with a series of doomsday warnings, press conferences, and hearings. Claims went beyond the absurd: Senator Gore even asserted that ozone depletion could be a factor in the spread of AIDS. Everyone—Environmental Protection Agency chief William Reilly, the editorials of major national newspapers, the DuPont Company, and an array of environmentalist groups—was calling for an accelerated phaseout of CFCs. On Feb. 11, 1992, President Bush announced a ban on CFCs by 1995, the schedule suggested by the Natural Resources Defense Council.

We have not been able to examine the impact of these new mandates in detail; the data in this chapter are based on the original Montreal Protocol schedule. What can be said, however, is that the new mandates will have even worse consequences than those reported here.

The State Department official who negotiated the Montreal Protocol, Richard Elliot Benedick, an admitted Malthusian, summarizes the shock effect of the Montreal Protocol chemical bans on the modern world economy in his book *Ozone Diplomacy:*

> The Montreal Protocol on Substances That Deplete the Ozone Layer mandated significant reductions in the use of several extremely useful chemicals. ... By their action, the signatory countries sounded the death knell for an important part of the international chemical industry, with implications for billions of dollars in investments and hundreds of thousands of jobs in related sectors. The protocol did not simply prescribe limits on these chemicals based on "best available

technology," which had been a traditional way of reconciling environmental goals with economic interests. Rather, the negotiators established target dates for replacing products that had become synonymous with modern standards of living, even though the requisite technologies did not yet exist. ...

At the time of the negotiations and signing, no measurable evidence of damage existed. Thus, unlike environmental agreements of the past, the treaty was not a response to harmful developments or events but rather a preventive action on a global scale [pp. 1–2].

The consequences and costs for the world economy of the ban on CFCs and related chemicals cannot be presented here in full. Perhaps one example—how the ban will affect air-conditioning—will suffice to give an idea of the scope of the disaster. As a result of the ban on CFCs, the air-conditioning systems in 100,000 large buildings in the United States and approximately 430,000 large buildings elsewhere will have to be torn apart and replaced. Air conditioning systems are a necessity for modern buildings that are constructed without movable windows and cross-ventilation, in many cases because of "energy-conservation" measures imposed by the environmentalists. Their air-conditioning systems will have to be replaced virtually immediately, because they leak Freon and have to be recharged frequently, unlike refrigerators, in which CFCs can remain sealed inside the system for the lifetime of the unit.

How much will this cost? If each replacement job runs $1 million—a very conservative figure—that would total $530 billion. To replace the air-conditioning systems of the Sears Tower in Chicago or the World Trade Center twin towers in New York would cost much more.

In the refrigeration industry, the ban on CFCs creates a different kind of nightmare. Refrigerator manufacturers not only have to come up with new refrigerator designs, but also have to scrap their entire production lines and machine tools. This all has to happen by 1993.

What About Replacements?

There is a fierce race among the giant chemical corporations to develop chemicals that will replace CFCs. Du Pont Company,

Imperial Chemicals, Allied Signal, Hoechst, and another 10 chemical giants have each spent hundreds of millions of dollars—$3 billion total—in the search. The expenses are high, but so is the payoff: The company that patents a replacement for CFCs stands to make tens of billions of dollars in profits.

However, a replacement has yet to be found. As the deadline for the total ban on CFCs approaches, there is hysteria in the refrigeration industry. This is reflected in the ASHRAE *Journal* mentioned above. Nearly the entire February 1991 issue is dedicated to articles describing the fallacies in the widespread beliefs about replacing CFCs.

Let's look at the leading contender to replace CFCs, Du Pont Company's HFC–134a, also called Suva. Suva is incompatible with all existing lubricants. A new special-purpose (and very expensive) lubricant has just been synthesized, which works for a short period of time. But even tiny traces of the old lubricants are enough to begin a chain reaction of corrosion within the refrigerating machinery in which Suva is used to replace CFCs. Suva is extremely corrosive to metal parts and less energy efficient than CFC–12. To top it off, Suva costs 10 to 30 times more than the CFCs it replaces—although it is the cheapest substitute yet developed.

Theodore Atwood, a senior research engineer with the Buffalo Research Laboratory of Allied Signal Inc., makes several very important points about the CFC replacement problem, in the ASHRAE *Journal.* "The current phase-down and anticipated phaseout of CFCs 11, 12, 113, 114 and 115, coupled with uncertainty over the long-term acceptability of R–22, has created uncertainty and confusion," Atwood writes. Some people "have over-optimistically assumed that simple quick-fixes would emerge to dispel all problems. . . [and] the uncertainties of this transition period have provided a fertile breeding ground for many ideas, claims, and suggestions ranging from 'off the wall' on through simple naivete. Add in a sprinkling of the inevitable opportunism and we have the recipe for rampant confusion. Unfortunately the magnitude of the task of reconfiguring product lines and entire industries to comply with regulatory timetables leaves little latitude for leisurely contemplation. . ." (p. 30).

Atwood cites and rebuts nine fallacies about CFC replacement. One of these, that "there must be an abundance of suitable candidate fluids just awaiting discovery and development, [is] grossly in error," he says:

Chemistry, like all science, is rigidly bounded by what nature will allow; it cannot create miracles on demand. Unfortunately, we are already scraping the bottom of the refrigerant barrel.

Despite extensive worldwide efforts spanning many years (dating back before the CFC/ozone hypothesis), no freshly discovered species of promising new refrigerants has emerged. While it is true that significant numbers of new compounds are annually added to the world's chemical roster, there is no corresponding increase in potential new refrigerant molecules. Today's possibilities for synthesizing new chemical compounds are primarily complex molecules of potential interest as pharmaceuticals, polymers and other applications that have no utility as refrigerants [p. 31].

Fallacy number three, according to Atwood, is that "the public should anticipate universal availability of 'drop-in' replacements for the refrigerants mandated for phase-down to sustain the systems and devices currently in the field throughout their normal lifespans." In other words, people expect to keep on enjoying the convenience of refrigerators and air conditioners with new chemical refrigerants, but it is unlikely that they will be able to do so. Furthermore, he says: "The refrigerant is the fundamental component around which each system is designed and fine-tuned. It imposes specific requirements in the choice of system materials. It dictates size and design of individual system components and of the complete system itself."

In other words, it is very unlikely that your old refrigerator will work without CFC refrigerants.

Atwood elaborates:

A replacement refrigerant can never possess identical characteristics unless it is, in fact, the identical molecule. Replacement molecules, at best, may approximate some of the important characteristics of the original fluid. This could be in terms of modified operating conditions, altered performance and/or incompatibility with existing system components and materials such as motor windings, seals or lubricants. There are also secondary problems, such as retraining all refrigerator repairmen, and maintaining duplicate inventories of repair parts during the transition years.

Even those who can afford to buy a new refrigerator cooled by Du Pont Company's Suva (HFC–134a) won't be out of the woods financially. Because the new coolant destroys lubricants, the compressor of the new refrigerator will grind itself into destruction, refrigeration industry experts expect refrigerators using HFC–134a to last between four and seven years—compared to eighteen or more years for today's conventional refrigerators. (Interestingly, the machinery of a standard refrigerator will operate virtually forever; what breaks down is wiring, insulation, door gaskets, cold controls, etc.) With Suva, those who can pay the steep price of a new refrigerator will be forced to do so every four to seven years.

Refrigeration experts cannot agree on whether HFC–134a will work or not. Some experts say that the lubricant problem has been solved and production of new refrigerators can start in 1995. Others say a solution is still to be found. However, a lubricant that can stand up to Suva may cost as much as the refrigerant, boosting the cost of new refrigerators even higher. The problem does not end there. Suva also reacts with the desiccant that is used to keep water out of the refrigerant, and no solution to this problem is yet in sight. And even if a lubricant and a desiccant compatible with Suva are found, the problems of corrosibility and possible toxicity to human beings remain. There is reason to believe that Suva qualifies under U.S. government definitions as a "cancer-causing substance." It is still too early to know. In addition, a large percentage of the Earth's population will soon be exposed to another potentially lethal risk: HFC–123, a replacement in foam-blowing applications, has already proven to be highly poisonous when subjected to heat.

Energy Inefficiency

Then there is the issue of energy. The banning of CFCs, in one stroke, has succeeded in taking back the gains in energy conservation achieved over the past two decades by use of more and improved insulation. CFCs, the most efficient blowing agents for insulation, can no longer be used. The amount of energy that will be wasted as a result was calculated in a study on the CFC ban conducted at Oak Ridge National Laboratory before the 1990 London conference. The updated Montreal Protocol endorsed at the London conference was even more extreme than the "worst-case" scenario projected by the study.

That "worst-case" analysis in the Oak Ridge study showed that the penalty of the CFCs ban, measured in units of energy wasted, will be as much as 2.18 quadrillion Btus per year—in the United States alone. To imagine an amount of energy this immense, note that it takes 170 million barrels of oil to produce 1 quadrillion Btus of energy. This means the banning of CFCs will result in the burning of an additional 370 million barrels of oil per year (or equivalent) in the United States, more oil than is now used annually in most countries.

The increased energy use for electricity-powered building equipment alone shows a huge waste of energy. To operate the same equipment in the buildings would require an increased demand of more than 94 billion kwh/yr. This equipment includes refrigerators and freezers, cold-drink vending machines, drinking fountains, retail refrigeration, and water heaters. Most countries of the world don't *produce* this much electricity, much less waste it.

The United States does not have this energy to waste. Under present conditions, the nation's electricity grid is being strained to its limits; brownouts are becoming more and more frequent in many areas. Thanks to the environmentalist campaign against electricity generation, nuclear power plants are no longer being built and other baseload generating stations are not planned. Furthermore, under the Clean Air Act, it will be impossible to build new fossil-fueled power plants, and many of those now operating will be shut down. The ban on CFCs will be the climax of the 20-year environmentalist drive to wreck the electric grid of the United States and the rest of the world.

It is not just refrigeration of food that is under threat. What about the medicines and blood supplies that require refrigeration? Ralph Jaeggli, an engineer with many years of experience in the refrigeration industry, has warned that the ban on CFCs will have a serious effect on the world blood supply. According to Jaeggli, CFCs are the only safe refrigerant for the blood supply in a situation where even the most minute contamination from the refrigerant can have severe toxic effects on blood in storage. "The vital emergency blood supply at both normal temperature (4°C) and long-term frozen storage (−85°C) will be severely affected" by the ban unless suitable refrigerants and lubricants are invented, Jaeggli told the authors. "The industry is working feverishly on alternate refrigerants, but the task is absolutely monumental, involving new compressor designs, new lubricating oils, toxicity and corrosion testing." So far, no new methods of refrigerating blood have been discovered.

Jaeggli said he was astonished at the foolishness of the decision to ban CFCs. "The economic impact of the ruinous taxation on CFCs now in place and the eventual banning of CFC production worldwide will have stupefying economic and social repercussions."

Goodbye, Air Conditioning

The environmentalists claim that air conditioners are a luxury, revealing their lack of knowledge of human health, not to mention comfort. On hot days, air conditioners preserve life. Heat stroke was until recently a major cause of death for elderly people in the United States. Air conditioning has had a marked effect in reducing the death toll from heat stroke and related medical complications.

Under the present ban on CFCs, automobile air conditioners will be severely affected. Some 95 million cars in the United States are equipped with air conditioners, while in the rest of the world there are perhaps another 60 million units. General Motors, Ford, and Nissan corporations have announced that they will use Du Pont's patented HFC–134a in the air-conditioning systems of their 1993 model cars.

What does the changeover mean for automobiles? First, as mentioned above, Suva or HFC–134a is 30 times more expensive than Freon–12. It also is much less efficient, necessitating a larger and heavier compressor in the auto. This means the entire engine compartment, already overcrowded with yards of vacuum hoses and other antipollution devices, will have to be redesigned. Not only is HFC–134a much more corrosive, but also it destroys the lubricants in the compressor.

An experienced air-conditioning engineer reports that he has surveyed the results of extensive testing done on prototypes of the new HFC–134a-cooled systems and talked to the engineers at GM and Ford. They have already resigned themselves to the fact that their new compressors will last three years at most, rather than the ten years a CFC-cooled system will run. Repair costs on the new systems will also be exorbitant. A simple repair for present air-conditioning systems cost between $100 and $150 in 1991. For a 1993 model car using HFC–134a, a simple repair will cost $200 to $300, and the air-conditioning system may have to be replaced every three years at a cost of more than $1,000 Once the ban is in effect, the poor fellow bringing in his 1992 car for a simple recharge will have to pay some $700 to have the car's air conditioning system rebuilt.

And, as if that were not enough, gas mileage is going to go down with the new air-conditioning units.

How do we know so much already about HFC–134a? Experimental quantities of the new refrigerant have been available for some time, showing the replacement to be a real dud. A report on the results of testing appears in the March 1990 issue of *The Accumulator,* the newsletter of the Automotive Air Group. The editorial introduces the report, written by member Randy Scott, stating: "The impact R[efrigerant] 134a will have on our industry is real and imminent and powerful. Nothing of this magnitude has ever come along before. We will have to face up to the challenge very soon to find new solutions to old problems plus new problems not yet identified. It is inevitable; R12 [Freon–12] systems will be phased out sooner than most of us expected." The editorial warns further that, "the extent of incompatibility as reported in detail by Scott is unbelievable."

Scott reports:

R[efrigerant]–134a requires new type(s) of lubricant. It cannot be introduced as a mixture into systems that have *or have ever had* R12 present (even in traces, including after flushing) without extreme damage resulting. Condenser capacity/function(s) must increase for use with R134a in order to reduce head pressure inherent in R134a (especially in low-speed and city driving). New receiver/driers will probably be very similar in appearance to driers currently used for R12. *Do not use a R12 drier in an R134a system!* Breakdown of desiccant will cause severe damage [p. 5].

Some words of advice from Scott to automobile mechanics indicate that the average service station will be required to spend thousands of dollars on new equipment, to deal with cars using Du Pont's new refrigerant: "It will be most imperative that a 'fresh' set of gauges, vacuum pump, and recycling equipment be used initially on and thereafter only on R134a systems. Use a separate set used only on R12 systems. Even the amounts in the gauge hoses, etc. can cause severe problems when allowed to mix." In other words, every auto shop will have to purchase a new set of tools and equipment to repair air conditioners after the CFCs ban, and these tools must never come into contact with a CFC; if they do and are then used with HFC–134a, they will destroy the unit.

Furthermore, Scott says, "Leak detectors, both flame and electronic, *will not* detect R134a. There is no consumer-practical (due to price, mostly) detector available now." How can a leak in the air-conditioning systems of a car be fixed if the mechanic can't find it?

We have, so far, looked at the cost to society of a CFCs ban. Will there be benefits? Yes, but for whom? Take one example: Within two to three years there will be 130 million automobiles in the United States using CFC-based air-conditioning systems. Replacing these units with the new "ozone-friendly" Suva (at $700 a pop), will create a market of $91 billion. Not bad for a completely unnecessary change.

The Green Gestapo

Another frightening aspect of the Montreal Protocol and the future regulations planned is the nature of the enforcement arm. It is not an exaggeration to say that America's industry and agriculture are now living under the rule of an environmental gestapo. Speaking on the day the London treaty banning CFCs was signed, Environmental Protection Administration chief William K. Reilly announced that the Justice Department had begun a forceful campaign against violators of existing CFCs-control regulations in the United States. Reilly, who headed the U.S. delegation to the London conference, told the press that Justice had sued five importers of CFCs, alleging violations of the Clean Air Act and an obscure EPA rule restricting imports of these substances.

On June 29, 1990, Unitor Ships Service, Inc. of Long Beach, California; Fehr Brothers, Inc. of New York; and three other companies were accused of having imported CFCs into the United States without obtaining permits from the EPA. The permit requirement is the result of a Jan. 1, 1989, rule governing production and import of CFCs. None of the companies had been given prior notice of the rule.

Paul Berg, president of Unitor Ships Service, told the *Los Angeles Times* that he was "rather upset" about the suit, "because we were advised wrongly" by the EPA's Seattle office. According to the

Times, Unitor in 1989 "responded to a cruise ship's emergency call for CFC–11 by transferring 1,270 kilograms of the coolant from its Vancouver office—after first checking with the EPA, Berg said. He added that Unitor exported a similar amount of the substance to Canada when it learned of EPA's objection. He said EPA has proposed a settlement, the amount of which he would not disclose, and his company has accepted. 'We don't have the resources to fight the government,' he said. If it lost the case in court, Berg said, it could be fined as much as $25,000 for each kilogram of CFC that it imported."

In New York, Fehr Brothers, Inc. immediately settled its case, agreeing to pay a $101,935 penalty. A company spokesman told *Executive Intelligence Review* magazine that the company had never heard of the regulation under which it had been indicted, but that the regulations were too complicated to permit an effective defense. Fehr Brothers was accused by the EPA of importing 192,000 kilograms of CFC–113, a cleaning solvent, without requesting the EPA's permission. Under the EPA regulation, the company could have been fined as much as $4.8 billion for the violation, had it not agreed to settle so quickly. Fehr Brothers now intends to leave the CFCs business altogether.

The Fehr Brothers settlement has given the environmental gestapo the precedent for prosecuting other importers, producers, and users of CFCs. Otto G. Obermaier, U.S. Attorney for the Southern District of New York, announced that the Fehr Brothers settlement was the first proposed consent decree in the nation enforcing the new regulations. He called the settlement "a first and important step in enforcing the environmental laws protecting stratospheric ozone." James M. Strock, EPA assistant administrator for enforcement, said, "This case demonstrates EPA's commitment to vigorous enforcement of the provisions of the Montreal Protocol. EPA will pursue violators of the regulations on stratospheric ozone to the full extent of the law."

The next step will be long jail terms for individuals who violate the absurd CFCs regulations. The Clean Air Act of 1990 has a section mandating jail terms of one year or longer, not only for individuals who release insignificant amounts of CFCs into the air, but also for those who provide technologies to Third World nations to manufacture CFCs. Is this what our Founding Fathers intended for the United States of America to become?

References

Theodore Atwood, 1991. "Refrigerants of the Future: Facts and Fallacies," *ASHRAE Journal,* Vol. 33 (February), No. 2, pp. 30–35.

Richard Elliot Benedick, 1991. *Ozone Diplomacy: New Directions in Safeguarding the Planet.* Cambridge, Mass: Harvard University Press.

D.W. Kaminski, 1988."Refrigeration and Worldwide Food Economy." Presentation at International Refrigeration Conference of the International Refrigeration Institute, Paris.

Dixy Lee Ray and Lou Guzzo, 1990. *Trashing the Planet: How Science Can Help Us Deal with Acid Rain, Depletion of the Ozone, and Nuclear Waste (Among Other Things).* Washington, D.C.: Regnery Gateway.

Randy Scott, 1990. *The Accumulator.* Automotive Air Group (March), p. 5.

Michael Weisskopf, "Study Finds CFC Alternatives More Damaging than Believed. Faster Phaseout to Protect Ozone Layer Supported," *The Washington Post* (Feb. 23, 1992), p. A3.

8

No More Chemicals?

Most people have heard about the ban on CFCs. Many do not know, however, that the Montreal Protocol also bans the production and use of many other useful chemical compounds that contain either bromine or chlorine atoms in their chemical structure. These include the halons, methyl chloroform, and carbon tetrachloride—each of which plays a critical role in modern industrial society.

Take the case of halons, a class of chemicals that is rarely mentioned in news reports on the ban. Halons are extremely useful chemicals that save thousands of human lives every year. Halons, chlorofluorocarbons that contain a bromine atom in their chemical structure, are a special branch of the CFC family. The bromine atom gives halons extraordinary properties in extinguishing fires and suppressing explosions; there is no other chemical known that can extinguish fires as quickly and effectively. Furthermore, halons are nontoxic, noncorrosive, and not damaging to electronics equipment. The toxicity of halons is so low that they can put out a raging fire without harming anyone present. Carbon dioxide, by contrast, which is also effective in fighting fires, suffocates people and animals. Other fire-fighting chemicals extinguish fires without harming people, but destroy electronics and computer equipment.

The only drawback of halons is that they are expensive, $6 per pound for Halon–1301, for example. They are generally used only in places where fire is a great danger, such as aircraft, hospitals, pipelines, ships, submarines, tanks, personnel carriers, the Strategic Air Command, missile silos, and the control rooms of nuclear power plants.

An article in *Risk Management Magazine,* September 1980, headlined "Fire Protection Saves Lives," vividly describes the importance of halons in fighting fires:

> Two years ago, after the 1,500 ton Mexican tugboat *Ballena* was seriously damaged by fire, the owners decided she needed a more modern fire protection system. The tug's original two-cylinder (100-pound) carbon dioxide reel and hose arrangement was destroyed early in the blaze.... The decision probably saved the 11-year old tug from even greater loss recently, when her crankcase blew up while she was working in the Pacific Ocean. Debris from the explosion severed the main diesel fuel line, starting a fire in the engine room. The fire was suppressed within 10 seconds by a system designed by Chemetron Fire Systems. ... The fire was confined to the engine room area.
>
> A.N. Tillet, the surveyor representing the tugboat's owners and operators, said, "The fire which ignited both the crankcase and oil in the bilges below was of such proportions that the crew felt no system could save the vessel. However, the Chemetron/Halon system performed perfectly. Operating staff were able to continue their efforts in the space, the fire was quickly controlled, and the vessel suffered only minor fire damage.
>
> J.W. Page, director, systems department for San Diego Fire Equipment Co., who supervised the installation, said that the fast reaction of Chemetron's system and Halon 1301 ... probably saved the lives of two men injured in the blast. Both men were trapped in the engine room and rescued after the fire was extinguished. The tug is also back at work [p. 54].

The article's conclusion? "Halon, bromotrifluoromethane, is capable of effectively suppressing a fire in a hazardous enclosure without endangering human life."

The military is another major user of halons. The *Army Times,* Jan. 22, 1990, described the Army's consternation over the possibility that these chemicals will no longer be available:

> The Army is struggling to prevent extinction of the one chemical that can protect its tanks' crews and expensive internal equipment from fire caused by armor-piercing antitank weapon explosions.
>
> Stemming from its concern over the eroding ozone layer, Congress this spring will debate more than 12 bills that would ban the production and use of Halon 1301, a gaseous substance deployed on more than 20,000 Army combat vehicles, including the M1A1 Abrams main battle tank and M2/3 Bradley Infantry Fighting Vehicle.
>
> The gas is deployed widely because it extinguishes fires almost instantly and does not harm crewmen or short out sophisticated electronic equipment or corrode machinery. . . . In 1969, the service selected halon over conventional fire extinguishers such as automatic sprinklers, carbon dioxide, dry chemicals, or foam.
>
> Once a fire erupts inside the tank crew compartment, the halon is automatically released and mixes with chemicals in the fire. The gas kills the fire in less than one-quarter of a second, sources say. Its success has prompted NASA and the Federal Aviation Administration to use halon to stave off fires in the space shuttle and commercial aircraft [p. 23].

According to Carmen DiGiandomenico, who supervises fire protection for the Army Materiel Command, "The Army is faced with a situation where industry and academics all say that we should not get our hopes up high that there will be an alternative. . . . There is no glimmer of hope."

One of the most extraordinary properties of halons is that they can put out an explosion in process, which no other chemical can do. There is an extraordinary description of this in *Aerosol Age,* April 1990, by technical editor Montfort A. Johnsen:

> Halons are used in the aerosol industry, both in fire extinguishing spray cans and for explosion suppression in gas houses. Their uses are linked to the preservation of property, but more importantly, to life-safety as well. When talking to

people in an aerosol plant it is not difficult to find one or two gas house operators or engineers who have had the unforgettable, momentary panic of seeing a fireball zoom into existence in front of their eyes, only to be extinguished in a roaring, twinkling of milliseconds by high-pressure halon gas. In the aftermath, there is always a time of reflection, of comparing the slightly reddened skin, some singed hair and body bruises with what could have happened if the fireball had been allowed to balloon out to 8 to 10 feet (3 m) across, before exploding outward, through a blow-out wall panel or ceiling hatch of the gas-house [p. 36].

Johnsen ends asking a most important question: "How many of our industry people have been saved because of Halon installations? A hundred?" It is more dramatic to put the question in reverse: How many will now die because of the ban on halons?

The lives of sailors in the Navy depend on halons perhaps more than anyone else. TV viewers in early 1991 saw news clips of Russian submarines surfacing with clouds of smoke and fire billowing from the open hatches and Soviet sailors frantically calling for help. Many more Soviet submarines have sunk without our knowledge. Why do we see Russian submarines on fire but not American submarines? The reason is quite simple: American submarines use Halon–1301, while most Soviet submarines do not because the U.S.S.R. has no factories that can produce Halon–1301. Fires frequently occur on board American submarines and surface ships, but casualties are very low because of halon-equipped fire-extinguishing systems.

A Navy officer who is greatly distressed about the ban on halons told us that if halons had been in use in April 1989 in the turret of the battleship *Iowa* when the detonation charges exploded, most of the sailors who died in that turret would still be alive today. The officer estimated that with a ban on halons, two or three warships per year would suffer extensive damage from fire, and submarines would face a much greater danger of being lost at sea. The loss of a submarine due to fire every couple of years would not be surprising, he said, and the number of sailors who would lose their lives could be in the hundreds.

Another consequence of a ban on halons would make professional environmentalists happy: The oil production complex on Alaska's North Slope will have to shut down. Halons are absolutely essential to companies producing oil and gas on the North Slope. They are used as an explosion suppressant in production facilities and for both fire and explosion protection along the Alaskan pipeline. Other fire-fighting sytems freeze in the low temperatures of the Arctic or would kill workers if they ever had to be used. Without halons, North Slope production will cease.

This brings us to the point of how much it will cost to ban halons. The cost involved concerns what halons protect: the most valuable equipment in modern society. What would happen if the U.S. telephone interchanges were to burn down or the computer rooms in Wall Street or the Chicago Board of Trade were to catch on fire? Halons would snuff the fire in a quarter of a second, causing little or no damage. Other chemicals might put out the fires, but all the equipment and data would be destroyed as effectively as if the fire had burned the building down.

The Argument Against Halons

The halons ban is justified by the environmentalists on the basis of an alleged danger to stratospheric ozone. The contention is that halons can destroy 10 times as many ozone molecules as other CFCs supposedly can. Such claims, however, are as fraudulent as all the other claims of the ozone priesthood. There is very little halon produced, and even less is emitted into the atmosphere. Most halons remain safely stored in containers, waiting to be used to put out a fire, save lives, and protect property. Of the 25,000 tons of halons produced in 1988, more than 21,000 tons remain stored in their original containers. The rest were used in fire-fighting or equipment testing. When used to put out fires, the halons are immediately changed chemically into other compounds that have very short atmospheric lifetimes. They cannot possibly affect ozone.

The 25,000 tons of halons produced annually is 0.02 percent of the 1.2 million tons of annual CFCs production. The "killer" molecule in halons—the one that is supposed to gobble up ozone molecules in the stratosphere—is bromine. The total amount of bromine in the halons produced annually is 12,040 tons.

Let's humor the environmentalists and say that all the halons released from containers each year escape into the air and do not break down in the process of snuffing fires. The amount of bromine thus released is still minuscule in comparison to the amount

Figure 8.1
NATURAL SOURCES OF BROMINE
COMPARED TO BROMINE IN HALONS
(tons)

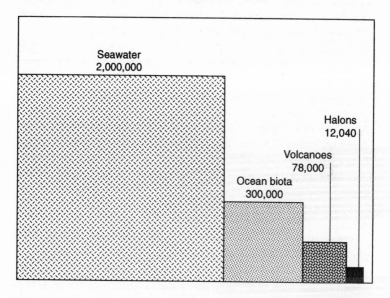

of bromine released from natural sources. As can be seen from Table 8.1, salt sprays from the ocean contribute 2 million tons of bromine to the atmosphere every year; more than 300,000 tons of methyl bromides are emitted by algae; and 78,000 tons are estimated to be blown out of volcanoes—even in years without great volcanic eruptions. In other words, the amount of bromine contained in halons is less than 0.6 percent of the amount of bromine spewed into the atmosphere every year from natural sources. If bromine *were* a threat to the ozone layer, there would not be an ozone layer left to speak of—Mother Nature would have already destroyed it.

The real concern is the thousands of people who will die every year in fires because of the ban on halons—murdered by the environmentalists, as effectively as if they had put a gun to their heads and pulled the trigger.

The United States already has one of the worst fire records of any industrialized nation—both in terms of numbers of fires and fatalities. For this reason the National Association of State Fire

Table 8.1
ATMOSPHERIC SOURCES OF BROMINE
(tons)

Sea salt	2,000,000
Ocean biota	300,000
Volcanoes	78,000
Total natural sources	**2,378,000**
Bromine in halons	12,040

Marshals and the National Volunteer Fire Council have passed resolutions stating unequivocally that halons must not be removed from the marketplace until acceptable substitutes are commercially available. The chairman of the Congressional Fire Services Caucus is also on record strongly opposing any attempt to remove this life-saving chemical from the market. Another fire-fighting expert emphasized his concern for the human lives that will be lost because of the ban on halons. The ban "is a cruel policy, with no regard for human life," he said.

These are not the only voices of sanity. One feisty midsize U.S. chemical company has been challenging Du Pont and the other giant chemical corporations that are behind the Montreal Protocol, insisting that halons do not pose a threat to the ozone layer.

Are there replacements for halons? Industry spokesmen say there are no replacements for Halon–1301, the chemical that can be used in confined quarters with no toxic effect. However, there is a fierce race to find replacements for Halon–1211, largely between Great Lakes Chemical Co. and the Du Pont Company. The new fire-fighting chemical invented by Du Pont to replace halons, HFC–123, proved to be incapable of stopping fires. Great Lakes' new fire-fighting chemical, Fire Master 100, has proven to be superior to Halon–1211 in fighting fires in extensive tests conducted by the Navy in Bufors, South Carolina. For ideological reasons, however, the EPA is refusing to allow the use of Fire Master 100, because it allegedly still has a slight ozone-depletion potential.

This has become a hot political issue for the military. According to industry sources, the Air Force is expected to announce a halt to all use of halons and the adoption of the Du Pont Company's HFC–123 as its new standard fire-fighting chemical. Since the Du Pont chemical has been shown to be a failure in field tests, the Air Force's position seems to be that it is willing to lose lives in order to score political points. Whether the U.S. Navy will follow the Air

Force is another matter. The Navy stands to lose not only men, but ships and submarines under a ban on halons.

The Medical Implications

Fire is not the only death threat that will result from a ban on CFCs and halons. Millions of individuals with respiratory problems depend on CFCs in their metered-dose inhalers.

Writing in *Aerosol Age,* August 1990, Terrance C. Coyne, M.D., chairman of the Pharmaceutical Aerosol CFC Coalition, states:

> Metered-dose inhalers (MDIs) have been popular since their introduction in the mid-1950s because they offer superior safety and efficacy, high levels of patient satisfaction and compliance, and reasonable therapy cost. Now, however, pharmaceutical companies are faced with a challenge because MDIs rely on mixtures of propellants that include chlorofluorocarbons (CFCs). . . .
>
> The amount of chlorofluorocarbons (CFCs) used as propellants in pharmaceutical aerosols is relatively insignificant. In the United States, they account for only 0.4 percent of total CFCs released into the atmosphere. At the same time, however, the medical need for MDIs is significant and growing.
>
> MDIs are used by approximately 24 million Americans who suffer from asthma, chronic obstructive pulmonary disease (COPD), and other lung diseases characterized by obstruction of airflow and shortness of breath. For some, MDIs are the only way to maintain the quality of their lives; for others, the devices are a lifeline. MDIs can arrest or prevent an episode of asthma or COPD by delivering a precise amount of active drug directly to the lungs in seconds. Without fast relief of bronchoconstriction, patients can suffocate and die.
>
> Nonaerosol alternatives do exist, such as dry powder inhalators (DPI), oral capsules or tablets, oral suspensions and syrups, and solutions for nebulizers. But often these alternatives cannot be used because a certain drug is not deliverable or effective in that form, or because the patient cannot tolerate the delivery form (as is the case with dry powder inhalation systems) or does not comply with the dosage [p. 24].

Many pharmaceutical companies and medical professionals have urged the United Nations Environment Program and other authors of the Montreal Protocol to make a special exemption for the medical uses of CFCs. With characteristic disregard for human suffering, the United Nations agency rejected the exemptions.

The environmentalists seem to think that a substitute for CFCs as medical propellants is in the bag. But, as Coyne has warned, "Each drug requires its own formulation, which can often include up to three [CFC] propellants. That can quadruple the challenge of developing a compatible formulation" (p. 25).

Pharmaceutical manufacturers have still another problem to deal with: They don't manufacture the propellants, the chemical companies do. These firms face a nightmarish task in replacing CFCs for the military, industry, and construction. It is likely that little more than a small portion of their efforts can be focused on medical applications. Furthermore, even if an alternative combination of propellants is found, it will be more than a decade before it becomes available to the public. According to Coyne, "For the approximately 15 currently marketed drugs, however, it will be at least 10 to 12 years before reformulation and regulatory approval can be accomplished" (p. 24).

The Recycling Boondoggle

The environmentalists push recycling as the solution to virtually every problem (except, of course, nuclear waste, which, by completing the nuclear fuel cycle and reprocessing would turn the waste into a resource). Recycling plays a big role in the Montreal Protocol guidelines, which assert that more than 30 percent of all CFCs still needed after a ban will come from recycling. Will this work?

The magazine of the Refrigeration Service Engineers Society, *Refrigeration Service and Contracting,* featured a cover story January 1990 on "the CFC Crisis," which examines the recycling issue concisely and competently. Here's what it reports:

Nine out of the thirteen companies [that produce refrigerant recovery and/or recycling equipment], claim to have equipment that can recycle CFC–12, CFC–500, CFC–502 and HCFC–22. Equipment for the recycling of CFC–11, CFC–113 and CFC–114 is in its early stages of development. . . .

According to ACCA [Air Conditioning Contractors of America] Executive Vice President James P. Norris, "If recovery/recycling of refrigerant is required by law, the demand for recovery/recycling equipment is expected to reach upwards of 325,000 units. Nationwide, sales of such equipment in 1988 totaled about 600 units.

"Since about 45,000 air conditioning contractors alone will be in need of this equipment, the market will undoubtedly experience tremendous growth problems and . . . contractors will face a shortage of recovery/recycling equipment," he said.

Norris also said that contractors will face a variety of other problems. "Because of the weight of most units, it would be extremely difficult for one service technician to carry most of the models currently available to a roof unit in order to recover refrigerant. Contractors may end up paying two technicians instead of one to do the job. This cost, of course will be passed on to the consumer" [p. 18].

Norris continued: "Processing time is also a problem. On the average, the equipment available is capable of recycling three pounds of refrigerant per minute. When large amounts of refrigerant are recovered, the time that it takes for a technician to recycle will result in higher labor costs for the consumer."

This statement, however, was challenged by automotive air-conditioning engineer Bob Holzknecht, who said: "The estimated recycling capacity of these machines is wrong. This is typical of the 'statistics' produced by a computer under the direction of a dedicated environmentalist. In the real world of refrigeration repair, the figure is near zero. Any active auto-air repairman can explain that usually when a component in a refrigeration circuit fails and must be replaced, the freon has already leaked out before he gets the job." According to Holzknecht, "it is the shortage of freon (lack of cooling) which alerts the owner he has a malfunction. Practically speaking, the only freon recovery possible is the temporary charge installed by the serviceman to assist him in pinpointing the leak(s) and determining the extent of necessary repairs." Holzknecht said: "Mandated recovery of freon from all sick refrigeration systems is like an elephant giving birth to a baby mouse. If one allows for incidental losses at each transfer and processing step, the net recovery is infinitesimal."

Using the little CFC that can be recycled may not be such a good idea either. According to Norris, "many air-conditioning contractors question the reliability of the recycling equipment currently available. The risk of damaging expensive air-conditioning equipment makes using recycled, possibly contaminated refrigerant, a huge gamble" (p. 18).

Other than damaging expensive cooling equipment, recycling CFCs can also be quite dangerous. "Filling metal vessels with high-pressure liquids or vapor is always risky," writes Anne M. Hayner in the same issue of *Refrigeration Service and Contracting*:

> But the assumption behind the current philosophy of recovery is that collected refrigerants will not only be pulled, processed, and put back into a system, but will also be transferred from system to system, transported from field to shop, or maybe even sent away to a larger facility for testing and reclamation.
>
> That means stored refrigerants will be subjected to multiple temperature changes and may be transported over state lines.
>
> That's not so dangerous if you are using a Department of Transportation (DOT)-approved refillable container with a safety release valve—and if you know what you're doing.
>
> But if the tank isn't and you don't, chances are you'll be caught in a situation that's not only dangerous, but also illegal [p. 26].

Roger Dumais, of Thermaflo Company, put it this way: "You will virtually have a bomb on your hands."

The Montreal Protocol also requires refrigeration equipment to be refilled with recycled Freons, which may have impurities that can destroy the equipment. According to Hayner, "While many reputable manufacturers offer products that recover and clean refrigerant, most units can't return refrigerant to original specifications ... cannot guarantee evacuation efficiency or the purity of the recycled refrigerant."

She reports:

> "You can make a machine that will pull refrigerant out of an air-conditioner or refrigerator," says Jim Lawless, Davco Manufacturing Co. "And you can make a machine that will clean it up, if you know what [contaminant] you're cleaning

up. The great difficulty is instrumentation to measure the contaminants in the refrigerant. And, after you've taken them out, [you need] to measure that you've taken them out."

Refrigerant purity levels can only be determined reliably with a gas chromatograph, with nuclear magnetic resonance, or by chemical analysis.

"Small-quantity recycling quality verification is tough to justify," says Du Pont's Ron Masterniak. "Lab analysis fees are very expensive; about $100 to $300 per sample. Can contractors afford to pay $300 to find out if the 50 lb. of refrigerant they've removed from a system is safe to put back? The answer, most likely, is 'no' " [p. 23].

Lab analysis fees are not the only expense involved in recycling CFCs. Manufacturers may be forced to purchase recycling units, if legislation now being heard on Capitol Hill is passed. Recycling units cost an average of $3,000, but can cost as much as $50,000, depending on which CFCs they recycle and whether they simply recover the CFCs or also recycle them. Regulations being implemented now by the Environmental Protection Agency mandate that every shop in the United States that repairs just one single air conditioner per year has to purchase a recycling unit. At least 375,000 recycling units will be required to comply with this mandate, and the number may be significantly higher. Using these figures as a baseline, the cost of the equipment to recycle CFCs will run somewhere between $975 million and $16 billion.

Even these figures do not include the costs of installation, filters, and servicing of the units. Also missing are the labor cost of operating the machine and the cost of operators' specialized training to qualify for mandated certification. Each of these nonproductive costs will be marked up and passed along as part of each repair job. No wonder manufacturers of CFC recycling units are lobbying for tougher laws to ban the production of new CFCs.

The End of Electronics?

Methyl chloroform is another chemical that will be banned under the London Treaty amendments to the Montreal Protocol. As with CFCs, methyl chloroform is not a glamorous chemical. It is there, it does its job, and 99 percent of people never heard of it. The same people, however, will be severely affected by its banning.

Methyl chloroform is used widely in the electronics, aerospace, automotive, defense, medical products, aerosol, coating, textile, and adhesive industries. It has become the solvent of choice for these industries in the United States, Europe, Japan, and elsewhere for straightforward reasons: It is virtually nonflammable and non-combustible, relatively nontoxic, and not photochemically reactive, therefore not contributing to urban air pollution in the lower atmosphere.

Even the professional environmentalists cannot do without methyl chloroform. When they turn on their stereos or their VCRs, or climb into their Mercedes Benzes, BMWs, or Volvos, they are using technologies that depend on methyl chloroform.

It is virtually impossible, in fact, to overstate the importance of methyl chloroform in our industrial economy. Paul A. Cammer, president of the Halogenated Solvents Industry Alliance, summarized its uses in testimony July 11, 1990, to a House subcommittee of the Committee on Science, Space, and Technology. We paraphrase from his testimony:

Electronics. Major electronics and aerospace companies depend on methyl chloroform to safely clean millions of high-technology printed circuit boards that are used in aerospace and defense applications, including communications, weather and military satellites, inertial guidance gyroscopes for aircraft and satellite guidance systems, ballistic and tactical missile guidance systems, and radar detection systems.

Methyl chloroform is important to the production of printed circuit boards in the manufacture of medical, business, and consumer electronics such as computer disc memory storage components, personal and mainframe computers, calculators, telefax machines, televisions, telephones, and stereos. It is also used in the manufacture and maintenance of electric and mechanical instrumentation of nuclear, hydroelectric, and chemical plants—instrumentation that is crucial to the safe and efficient operation of these plants.

Automotive. Methyl chloroform is critical in the automotive industry for cleaning (defluxing) hybrid circuits, particularly ceramic substrate electronics (that is, ignition modules, engine controllers, fuel sensors, and voltage regulators). It is used in the production of a wide variety of plastic and metal parts for the automotive industry, and it is critical in the installation and safe operation of automotive and aircraft braking systems.

Metals. Methyl chloroform is used where high-performance cleaning is essential to prepare surfaces for subsequent assembly. Clean surfaces are necessary when parts need to be joined securely without slippage or creep that could lead to failure of products in use and to permit application of protective coatings.

Aerospace. The aerospace industry is particularly dependent on the use of methyl chloroform because of its heavy reliance on aluminum, which cannot be easily cleaned with aqueous/alkaline systems. Methyl chloroform is essential to clean the Space Shuttle booster casings used to launch the NASA Space Shuttles, and it is also essential in the production of aircraft sheet-metal assemblies, and the nondestructive detection of fatigue cracks in commercial and military aircraft wings.

Construction. Methyl chloroform is essential to clean the ends of wire rope used in bridges and on cranes and to catch jets landing on aircraft carriers, where reliability is crucial.

Medical. Methyl chloroform is the only compound approved by the Food and Drug Administration for the purification of hollow fibers used to provide oxygen to the blood during heart and kidney surgery. It is also used in the production of syringes, cardiac pacemakers, and artificial joints, limbs, and implants.

Adhesives. Methyl chloroform is essential in the formulation of nonflammable adhesives, allowing companies to comply with smog control regulations and to avoid the high product liability premiums associated with flammable products.

Aerosols. Methyl chloroform is present in aerosol insecticides to reduce or eliminate flammability. It is a preferred solvent in wasp spray applications because of its instant knockdown ability, low toxicity, and safety of use. For those unfortunate individuals who may be allergic to the sting of a wasp, this product is literally life-saving.

215

Textiles. Methyl chloroform is used commonly in the textile industry to safely and effectively remove oils and other difficult stains from cloth in the production of broadcloth and other textiles.

Cammer also told the subcommittee that more than 73,000 manufacturing companies employing about 3,000,000 U.S. workers rely on methyl chloroform—23.5 percent of all U.S. manufacturing production employees. In 1988, these firms accounted for $295 billion of value added, or 25.1 percent of total gross manufacturing product. The percentage of U.S. industrial establishments using methyl chloroform ranges from a low of 5 percent in textiles and plastics manufacturing to more than 55 percent in electronics, automobiles, and other transportation equipment manufacturers.

Without methyl chloroform, many small businesses will drop certain products or shut down, because they have no suitable substitutes or because it is too costly to redesign and retrofit their production processes. Other businesses will be forced to risk reduced quality of output—a key competitive factor, especially in high-technology industries. Virtually all will have to undergo expensive redesign and shift to higher-cost replacements for methyl chloroform, if there are replacements available. Moreover, in the United States, most of the alternatives proposed by the environmentalists as substitutes for methyl chloroform are being outlawed by the Clean Air Bill for allegedly contributing to urban smog.

References

Terrance C. Coyne, 1990. "A Status Report: CFCs in Pharmaceutical Aerosols," *Aerosol Age,* Vol. 35 (August), No. 8, pp. 24–25.

"Environmentalists Challenge Use of Fire-Extinguishing Halon Gas," 1990. *Army Times* (Jan. 22.)

"Fire Protection System Saves Lives," 1980. *Risk Management* (September), pp. 54–55.

Anne M. Hayner, 1990. "Measuring Recovery/Recycling Reliability," *Refrigeration Service and Contracting* (January), pp. 24–26.

Montfort A. Johnsen, 1990. "The Halons: Present Uses—Future Options," *Aerosol Age* (April), pp. 36–39.

"What's On the Recovery/Recycling Market?" 1990. *Refrigeration Service and Contracting* (January), pp. 18–19.

9

The Corporate Environmentalists

The last two chapters have documented that the ban on CFCs will cost the lives of 20 to 40 million human beings every year and drain the world economy of $3 trillion to $5 trillion by the year 2005. This chapter will examine the other side of the issue: *Cui bono?* Who benefits, and what kinds of benefits will they reap? Who will bank the enormous profits that stand to be made from marketing the new chemicals and equipment required by the ban on CFCs?

And, just as important, who will extend control over the world economy as a result of the ban on CFCs? The Montreal Protocol has established the kernel of the "New World Order" to which George Bush often refers. In this brave new world, a small elite of technocrats will dictate economic policies to all the nations of the world. Any nation that disobeys this new dictat will pay the consequences, as outlined in the Montreal Protocol: That nation will suffer a total economic embargo. In effect, a new world colonial regime is being set up under the guise of saving the Earth from man's pollution. Environmentalism is being used as the cover for the takeover of natural resources, the destruction of sovereign national governments, and the imposition of population control.

This is exactly the proposal made by Soviet leader Mikhail Gorbachev in December 1988 during his historic visit to the United Nations. In his speech to the United Nations, Gorbachev called for a world environmental order, run by an ecological security council operating within the United Nations and deploying its own military, with the power to enforce and oversee the implementation of environmental policies. This international green police would operate independently of sovereign national governments with the power to intervene militarily against nations accused of polluting the environment. Gorbachev's proposal received a warm welcome from the government of former British Prime Minister Margaret Thatcher and from the incoming Bush administration.

The creation of this green gestapo is much closer than generally realized. Such a supranational environmentalist institution is scheduled to emerge from the June 1992 United Nations Conference on Environment and Development, the so-called Earth Summit, to take place in Rio de Janeiro, Brazil. Government leaders and several thousand environmentalist activists will gather in Rio from across the globe to hammer out the details of a new Earth Charter, which, among other things, is intended to impose restrictions on the emission of gases into the atmosphere. Such a treaty implies the control of economic activity and growth worldwide. The model for the Earth Summit is the Montreal Protocol.

Were it to become widely known that the ozone depletion theory is a fraud, the New World Order and its Earth Charter would begin to fall apart.

The plans for this New World Order and its legislation are not secret. They are spelled out in great detail in the publications of the leading environmental groups. All of its major elements have been introduced to Congress as legislation, by former Representative Claudine Schneider, Senators Albert Gore and Timothy Wirth, and others.

This chapter will concentrate on the more easily understood aspect of this New World Order: money. (The philosophical aspect is covered in Chapters 10 and 11.) There is a lot of money to be made by banning CFCs and other useful chemicals. Environmentalism has become very fashionable in the business community. Major companies are rushing to put out "environmentally friendly products," and "recycling" is becoming a corporate religion. The stocks of companies considered "green" have zoomed on Wall Street, while the stocks of corporations considered "polluters" are being dumped.

The Du Pont Company is among the leaders of this new "corporate environmentalism." In fact, that term was coined by Du Pont chairman Edgar Woolard. The policy precedes him, however. According to Du Pont company insiders, the decision to support a full ban on CFCs was made in 1986, by Du Pont's new owners, Edgar and Charles Bronfman.

As described in greater detail below, the Bronfman family, which took controlling interest of the Du Pont Company in the early 1980s, made its fortune producing bootleg liquor that was smuggled into the United States during Prohibition. This is ironic, as the Montreal Protocol defines CFCs as *controlled substances,* whose use is regulated by law, just like cocaine, or, more to the point, like alcohol during Prohibition. This is a very profitable business indeed, especially if one controls the patented chemicals that will replace the controlled substance.

The European edition of the *The Wall Street Journal* on June 29, 1990 made the point very clear. Appearing on the day that the revisions of the Montreal Protocol were signed in London, the article reported: "An expected global agreement Friday to phase out many ozone-destroying chemicals will force an industry shake-out that may ultimately benefit the world's chemical giants." According to the *Journal,* "the accord will cause turmoil in the world chemical industry that only the strong will survive, industry officials say. In place of today's $2 billion-a-year world market for CFCs and halons, a new market for ozone-friendly chemicals will emerge. That new market will favor the chemical giants, which have the big labs and bulging treasuries needed to develop ozone-friendly substitutes. Global development costs are likely to exceed $4 billion—a sum only the industry powerhouses can easily afford."

The *Journal* continued: " 'There's going to be a radical shake-out in the market,' predicts Bridget Paterson, a product manager at Britain's Imperial Chemical Industries PLC. 'There won't be 32 suppliers anymore in the world; it will be six to ten.' "

The *Journal* article concludes, "The most likely survivors in an ozone-friendly market are the leaders today in CFC production: ICI; Du Pont Co. of Wilmington, Delaware; Hoechst AG of West Germany; Atochem SA of France; Allied-Signal Inc. of Buffalo, New York; and Showa Denko KK of Japan. . . ."

The *Journal* was on target. The ban on CFCs has effectively created one of the most closely knit cartels in the history of commerce. These gigantic chemical corporations have total control not only over the patented products but also over the means of production.

On June 21, 1990, Du Pont announced it would build production facilities worldwide to produce replacements for CFCs. Du Pont spokesmen told the press that the company intends to invest more than $1 billion over the next few years to take the lead in commercializing production of alternative refrigerants. Plants are planned for Corpus Christi, Texas; Louisville, Kentucky; Dordrecht, the Netherlands; and Chiba, Japan. They will become operational between 1992 and 1995. The facilities will be capable of producing more than 140 million pounds of CFC replacements annually, and the company claims it can supply most worldwide refrigeration needs through the end of the century.

What kind of profits will the chemical giants rake in?

As discussed in Chapter 7, the ban on CFCs will involve scrapping hundreds of millions of refrigerators, refrigerated transports, and cold storage rooms internationally. The profits involved in replacing all this equipment will be fabulous. The heart of all refrigeration systems, however, is the refrigerant chemicals. The profits generated from new chemicals will be truly staggering.

The giant chemical companies have already made more than $10 billion in revenues from the prohibition on CFCs. CFCs have become increasingly scarce since the signing of the Montreal Protocol in September 1987, and prices are now six to twenty times— depending on the product—what they were at the beginning of 1988. CFC–12, for example, the CFC most widely used in refrigerators and air conditioners, sold for 50 cents per pound in 1988; it now costs between $3 and $5 per pound. Experts in the refrigeration industry estimate that by 1995, when CFC production is scheduled to be reduced to 50 percent of what it was in 1986, the price for CFCs will have risen to between $15 and $25 per pound. It should be noted that the *cost* of production will be the same— approximately 50 cents per pound.

The U.S. government is part of the scam, through a tax on CFCs that became effective in January 1990. All existing supplies of CFCs in the United States and all future imports and production of CFCs are now taxed. In 1990, the tax was $1.37 per pound (more than three times the cost of production), and it will rise steeply, to almost $5 per pound by the end of the decade. The CFC tax will bring tens of billions of dollars into the U.S. Treasury, as part of the Bush administration's strategy to increase taxes under the cover of "saving the Earth."

The profits the chemical giants will continue to make on CFCs, however large, pale in comparison to the profits to be made from the chemicals that will replace CFCs.

The Corporate Environmentalists

Is the chemical industry a sleeping giant, unaware of the financial stakes in the ban on CFCs? The story tellers of the news media have spun many a yarn about heroic environmentalists waging a titanic struggle to save the Earth by forcing corporations to change their policies. These are nothing but fairy tales.

The truth is:

(1) The CFCs ban was pushed by the giant chemical corporations, led by Du Pont and Great Britain's Imperial Chemical Industries (ICI).

(2) As early as 1986, the banning of CFCs became a top priority for the U.S. State Department and intelligence services, as documented in Richard Benedick's book, *Ozone Diplomacy.*

(3) The environmental movement was founded and is paid for by the foundations that represent the accumulated wealth of American and European blueblood families, the same families that own and control most multinational corporations. Environmental groups receive more than $500 million a year in grants from the leading foundations, and more than $200 million a year in contributions from corporations. It is this funding that has enabled the environmentalist movement to obtain the enormous political clout it has today.

There is a widespread belief that environmentalists and corporations are antagonists and that the leading environmental groups are out to shut down industries worldwide. This is true to a certain extent, but should be qualified: There are huge profits to be made from environmental legislation—if one does not care about the welfare of human beings. There are tens of billions of dollars to be made on pollution-abatement equipment, new *patented* chemicals to replace those being banned, and merchandise that can be sold to duped consumers looking for environmentally friendly merchandise. Billions more can be made by taking land out of agricultural, industrial, or urban development. Land trusts,

221

for example, owned and controlled by America's leading blueblood families, represent immense wealth from real estate value and from tax breaks.

Most significant, small- and medium-sized businesses and industries will be the ones that go out of business, eliminating the competition. While major corporations can afford to hire legions of lawyers to defend themselves from environmental lawsuits and to follow the fast-changing regulations, smaller corporations cannot. In the late 1800s, America's railroad tycoons and robber barons obtained their enormous power by using the likes of Bat Masterson to kill the competition's railroad workers and destroy their tracks and locomotives. Today's robber barons are using the environmentalists.

Fortune magazine took a long look at the "corporate environmentalists" in its Feb. 12, 1990, issue. Under the headline "The Environment: Business Joins the New Crusade," *Fortune* described in great detail how certain corporations have adopted this as their strategy for the last decade of the century. " 'In the Nineties environmentalism will be the cutting edge of social reform, and absolutely the most important issue for business,' " Gary Miller, a public policy expert at Washington University in St. Louis, told the magazine (p. 44).

The corporation most cited for environmental activism is Du Pont. "Du Pont also sees business opportunities in environmental concern," wrote *Fortune*. "The company announced in early December the formation of a safety and environmental resources division to help industrial customers clean up toxic wastes. Management forecasts potential annual revenues of $1 billion from the new business by 2000." According to *Fortune,* "One of Edgar Woolard's first acts after becoming CEO of Du Pont in April [1989] . . . was to deliver a speech in London entitled—and calling for— 'Corporate Environmentalism' " (p. 47). Fortune adds: "Woolard now meets at least once a month with leading environmentalists, and his company is taking what seem dramatic steps demonstrating its concern" (p. 48).

Woolard stumps for green causes outside of the corporate boardroom as well. He has been the featured speaker at major environmental conferences over the past two years. In November 1989, Woolard addressed the Globescope Pacific Conference in Los Angeles, the largest environmental conference ever to gather in the United States.

On May 4, 1989, having just taken over at Du Pont, Woolard gave an extraordinary speech before the American Chamber of Commerce in London, where he revealed Du Pont's new strategy of corporate environmentalism. "As I look ahead to the challenges Du Pont faces in the next decade and beyond, I see one of our chief concerns being environmental stewardship. . . ." Woolard told the conference:

> We in industry have to develop a stronger awareness of ourselves as environmentalists. I am personally aware that as Du Pont's chief executive, I'm also Du Pont's chief environmentalist. . . .
>
> We should seek out those opportunities to align ourselves with the environmental community and demonstrate where environmental and industrial goals are compatible.
>
> In other words, I am calling for corporate environmentalism. . . .

Woolard's speech outlined Du Pont's concern over the CFCs issue as follows:

> Science and technology are marvelous things. No one knows this better than those of us who work at Du Pont where we have witnessed the many good things that science and engineering make possible in our lives.
>
> But the flip side is that we sometimes position ourselves on an environmental issue on the basis of available technical or scientific data alone. We have been too inclined to act as though public wishes and concerns matter less than the technical opinions of scientists and engineers. *But in fact public opinion must be dealt with regardless of the technical facts. . . .*
>
> We followed the CFC/ozone depletion issue since its introduction as a plausible, though untested, scientific theory first published by Molina and Rowland in 1974. As early as 1975 our then chairman Irving Shapiro stated that if there was credible scientific evidence of harm to the environment, we would cease manufacturing these materials.
>
> The scientific case was slow in building—some data refuting, some supporting the assertion. But during the next ten years it became apparent that a case for damage to the atm<

sphere could be made. *In 1986 we led industry support of international negotiations that resulted in the Montreal Protocol* with its provisions for cutbacks. . . .

What will those future actions be? Well, we cannot sit around and wait for events to drive us. We have developed a corporate agenda for environmental leadership for the next decade. . . .

Industry has a checkered past of successes and failures in environmental matters, and as a result, manufacturers have been painted many colors in recent years. That will have to change. In the future we will have to be seen as all one color. And that color had better be green [emphasis added].

It should be noted that green is also the color of the dollar bill.

The Road to Montreal

The international treaty banning CFCs was signed in Montreal in September 1987 and is known as the Montreal Protocol. The roots of this treaty go back to the 1978 ban on the use of CFCs as propellants in spray cans. How the treaty actually came about, however, is somewhat of a mystery because all the meetings that led to the protocol were held in secret.

What is clear, however, is that without the support of the Du Pont Company, there would have be no Montreal Protocol.

Until 1986, Du Pont Company was the world's most ardent defender of CFCs. Then, suddenly, it did a "bootlegger's turn," coming down on the side of the environmentalists and calling for a phaseout and banning of CFCs. Du Pont's betrayal took U.S. industry by surprise, preventing any organized opposition to the Montreal Protocol.

In discussing the events leading to the company's sudden reversal, Du Pont family members and former Du Pont employees pointed to the role of Edgar Bronfman. They told the authors that in 1980, when his takeover of the Du Pont Company was finalized, Bronfman began to force radical changes inside the company. One of the most dramatic changes, they assert, occurred at the end of 1985, when the fight against the ozone depletion theorists was taken out of the hands of the Freon division of the company and turned over to "senior corporate management." "The Canadians," they allege, were behind the decision.

224

Du Pont's Freon division was staffed and led by some of the world's most competent scientists and leading experts on CFCs. They strongly opposed a ban of the chemicals.

About the same time, sections of the U.S. government became involved in organizing worldwide for the prohibition of CFCs. The framework of what became the Montreal Protocol was laid in a secret meeting that took place in June 1986 in Leesburg, Virginia, attended by ministers of several governments. There, the U.S. State Department made the banning of CFCs a top priority of U.S. foreign policy.

More details of secret negotiations that led to the Montreal Protocol have been revealed in the 1991 book *Ozone Diplomacy*, written by Richard Elliot Benedick, chief U.S. negotiator of the protocol. Prior to his work in drafting the Montreal Protocol, Benedick was the head of the State Department's Office of Population. A rabid advocate of forced abortions and sterilizations as policies to control what he considers the threat of population growth in nations of darker-skinned people, Benedick quit the population post to protest President Reagan's refusal to ratify such population control policies at the 1984 Mexico Conference on Population. The State Department reassigned Benedick to environmental affairs. After the signing of the Montreal Protocol, Benedick was made the equivalent of U.S. ambassador to the World Wildlife Fund, a post he still holds.

In his book, Benedick reveals how important an element of U.S. foreign policy the CFCs issue had become. After the Leesburg meeting, he wrote,

The Department of State . . . designed and managed a multifaceted strategy to gain acceptance of the U.S. position by as many countries as possible. Over the next months, about 60 U.S. embassies were regularly provided with talking points explaining the rationale behind the U.S. proposals, as well as with scientific and policy updates. Embassies were instructed to engage their host governments in a continuous dialogue to inform, influence, and demonstrate flexibility. A constant stream of cables between Washington and the embassies enabled the State Department to keep abreast of subtle changes in foreign attitudes and to provide new information responsive to other governments' concerns [p. 55].

Even if governments were convinced to go along with the CFCs ban, the policy still had to be sold to the populations of their nations. The environmentalist propaganda machine was called into action. Starting in 1986, tens of millions of dollars poured into the bank accounts of leading environmental groups, in the form of grants from the major foundations. This money was earmarked to push scare stories about ozone depletion and global warming. The World Resources Institute, for example, received a $25 million grant from the MacArthur Foundation in 1987 to "inform" the public about the dangers of ozone depletion and global warming. The chairman of the foundation at the time was the late Thornton Bradshaw, at one time also chairman of NBC and RCA.

The propaganda campaign conducted by the lavishly funded environmental groups was paralleled by one of the greatest propaganda campaigns ever conducted by the U.S. Information Agency. Benedick describes it:

> The media were an integral element of the diplomatic strategy. The U.S. government undertook major efforts to reach out to foreign public opinion, especially in Europe and Japan, to counteract the previously unopposed influence of commercial interests. Senior U.S. officials and scientists gave speeches, press conferences, and radio and television interviews in numerous foreign capitals. Using the advanced Worldnet telecommunications satellite technology of the U.S. Information Agency, Robert Watson of NASA, who had coordinated the WMO/UNEP [World Meteorological Organization/ United Nations Environment Program] science assessment, and the chief U.S. negotiator [Benedick] appeared together in a series of live televised question-and-answer sessions involving influential foreign participants in over 20 capitals in Europe, Latin America, and Japan. These programs, which went on for more than a year, attracted considerable foreign television, radio and newspaper coverage [p. 56].

The Role of Britain's Chemical Giant

During this time, the British government dropped its opposition to a CFC ban.

What transformed former British Prime Minister Margaret Thatcher into an enthusiastic proponent of the ban? The powerful chemical interests of Great Britain.

Aside from the Du Pont Company, the major participant in the emerging world chemical cartel is Great Britain's Imperial Chemical Industries (ICI). The company, founded by Lord Alfred Mond in the last century, was the world's most powerful chemical concern during the heyday of the British Empire. The Empire's policy was to prevent the development of industry in any of its colonies. The colonies provided the raw materials to Britain, and British industries, such as ICI, then transformed these raw materials into finished goods, which were sold back at exorbitant prices to the colonies.

This setup allegedly ended after World War II, but in truth it remains in place, in no small part through the functioning of the British Commonwealth. International financial institutions, such as the World Bank and International Monetary Fund, took the place of the empire, and continue to force Third World countries to provide raw materials for Western industries, and the "cheap" labor for manufacturing. Instead of "colonialism," this system is now called "free trade." Any attempt by developing-sector nations to build industrial and manufacturing complexes is a threat to this order.

Tracing the events that led to the signing of the Montreal Protocol in 1987 and the London update to that protocol in 1990, we come across a very prominent player: Denys Henderson, chairman of Imperial Chemicals Industries. Henderson was Prime Minister Thatcher's special adviser on CFCs! Like Du Pont, ICI stands to make tens of billions of dollars in revenues from the ban on CFCs. ICI, which has a greater influence in Europe than Du Pont, played a critical role in "convincing" European governments to support the ban on CFCs.

ICI's future lies in the post-CFCs era, according to chairman Henderson. Addressing the company's annual shareholders meeting in April 1989, Henderson called for the total elimination of CFCs. He told his shareholders, "Our aim is to become the world's leading chemical company." Henderson announced that ICI was spending £100 million to find alternatives.

Two months before Henderson's speech, in February 1989, British Prime Minister Margaret Thatcher, with Henderson as her personal adviser, had hosted the first conference in London on "saving the ozone layer," the biggest eco-extravaganza to date on this theme. Representatives from 124 countries were in attendance, including ministers from 85 countries. The conference was described by some as an event where U.S., Canadian, and British delegations armtwisted the Third World nations reluctant to sign the Montreal Protocol. After the conference, a chemical analyst with a leading London stock brokerage told *Executive Intelligence Review* magazine: "There are billions of dollars at stake. ICI is positioning itself to corner an extremely lucrative market" (p. 15).

There is one more interesting aspect of the role of Imperial Chemical Industries in the banning of CFCs. One of the heirs of the ICI family fortune, Lord Peter Melchett, is the executive director in Great Britain of the radical environmentalist group Greenpeace. Lord Melchett is the grandson of ICI's founder Alfred Mond, the first Lord Melchett. Greenpeace, which has an annual income of more than $100 million worldwide, has been one of the leaders of the campaign against CFCs. As Greenpeace's top man in Europe, Lord Melchett has been repeatedly denounced for his ties to ICI. The British lord replies that he sold his ICI stocks, so he no longer has a conflict of interest. It remains a fact, however, that his family will make billions of dollars from the ban on CFCs.

The Du Pont Company and ICI are working together on the CFCs issue. They have a common front in terms of policy, but most important, they are setting up joint production facilities for HFC–134a and other patented replacement chemicals for CFCs. Furthermore, Du Pont and ICI have formed and lead an international consortium that is testing the alternatives to CFCs. Such an arrangement, which used to be illegal, will enable chosen companies to bring their replacements for CFCs onto the market much more quickly than the competition.

Du Pont and ICI have been buddies for decades. In the earlier part of the century, the two companies had a near-monopoly in chemicals worldwide. This cartel was so powerful that no prosecution on the basis of antitrust legislation succeeded until Sept. 28, 1951, when Du Pont and Remington Arms were convicted in U.S. federal court of conspiring with Imperial Chemicals Industries to divide the wartime markets in munitions, chemicals, and small arms, creating a worldwide monopoly. During the trial, evidence was produced of trade pacts for dividing the world's trade and sharing each other's patents.

The Bronfman Gang

Now, let us take a closer look at the family that has emerged in control of the global chemical cartel's flagship company: the Bronfmans.

The Bronfman family is best known to Americans through its ownership of Seagram's, the biggest liquor company in North America. Besides whiskey, however, the family empire includes banking, mining, oil, real estate, and, lately, chemicals.

In the early 1980s, the Bronfmans mounted a takeover of the Du Pont Company, which culminated with their almost complete control of the company's board of directors in 1986. As mentioned above, the Bronfmans developed the strategy of implementing a ban on CFCs at the same time the company developed patented replacements for the CFCs, which they could use to corner the world market.

That the Bronfmans would come up with such a strategy is not surprising. The family made most of its fortune during Prohibition, when the family patriarch, Sam Bronfman, became the leading provider of bootleg liquor to American organized-crime syndicates. It is estimated that more than 50 percent of the illegal liquor distributed by organized crime gangs in the United States during Prohibition came from the Bronfmans' Canadian distilleries.

Canadian author Peter C. Newman documents the violent rise of the Bronfmans, from two-bit whiskey producers in 1915 to the heads of the largest liquor empire in the world by 1936. Prior to Newman's book, *King of the Castle,* the only other author who had attempted a Bronfman biography was Terence Robertson. Newman describes what happened to Robertson:

> . . . Roderick Goodman of the *Toronto Daily Star*'s editorial department testified that on January 31, 1970, the author had telephoned him from a New York hotel room to explain that he had been commissioned to write the history of the Bronfman family but that he had "found out things they don't want me to write about." Graham Murray Caney, another *Star* editor, testified that Robertson had told him his life "had been threatened and we would know who was doing the threatening but that he would do the job himself." While he was still on the telephone, Caney had the call traced and alerted the New York Police Department. Detectives burst

into Terence Robertson's hotel room just minutes before he died of barbituate poisoning" [p. ix].

Newman points out that the Bronfmans had experience—of a sort—with chemicals. He interviewed a surviving Canadian gangster to learn about the manufacturing methods for Bronfman's rot gut. Newman quotes the Canadian:

"A Mounted Police officer who took part in a raid on the Bronfmans' Yorkton distillery once told me how they made the liquor," recalls John W. Mack. . . . "They had large galvanized vats which were lined with oak. Into these they poured distilled water and alcohol, to which was added some caramel and sulphuric acid. The theory was that the acid would attack the oaken lining until it was burnt out. This completed the 'aging process' while the caramel lent the required color. After a chemist had sampled the brew with an hydrometer to pronounce it safe, the taps were turned on and the bottles filled. . . . I remember one Saturday night buying a crock of so-called J. & T. Bell, which five of us consumed out on the prairie in a Model T. About an hour later all of us were paralyzed to a certain degree, and by Sunday noon the ends of my fingers were still numb" [p. 76].

Between 1920 and 1930, more than 34,000 Americans died of alcohol poisoning.

Another book, the bestseller *Dope, Inc.,* documents the role of the Bronfmans in the Prohibition-era gangland wars from Chicago to Little Italy in New York that left more than 2,000 gangsters and 500 policemen dead.

Exactly how the Du Pont Company was taken over by Sam's sons, Edgar and Charles Bronfman, still remains somewhat of a mystery. Many Du Pont family members are certain that it was an "inside-outside job," abetted by former Du Pont Chairman Irving Shapiro. It began in May 1981, when Dome Petroleum made a tender offer for 20 percent of the oil company Conoco. Dome was not interested in buying up Conoco, but wanted Conoco's Canadian operations. The response by investors to Dome's offer was overwhelming: 52 percent of Conoco's shares came in, instead of the 20 percent it had asked for. That was a clear indication that Conoco was susceptible to a takeover.

Carlos de Hoyos

Edgar Bronfman in 1978.

At that point, Edgar Bronfman jumped in. Seagrams had $2.3 billion in cash on hand, which was used to purchase the 32 percent of the Conoco stocks available. Conoco, on the advice of the notorious takeover artist Joe Flom, of Skadden, Arps, Slate, Meager & Flom, rejected a takeover offer by Bronfman, and worked out a deal with Dome Petroleum. That meant Conoco now had to find a white knight to save the company from a hostile takeover by Bronfman. That white knight ended up being the Du Pont Company.

The J.P. Morgan banking interests came to Du Pont, asking the chemical company to take over Conoco to prevent the Canadians from swallowing such an important American company. Inside the company, Irving Shapiro, who had recently retired as chairman of the Du Pont company, still held the powerful post of chairman of the finance committee. Shapiro convinced the board of directors to approve of the merger with Conoco. The announcement shocked Wall Street investors, because it was generally believed that there was no good reason why Du Pont should want to use its valuable capital to save Conoco from a hostile takeover.

Why would Shapiro propose such a course of action for Du Pont? It is important to note that Shapiro worked for many different interests. Aside from his Du Pont posts, Shapiro was a member of the law firm of Skadden, Arps, headed by the same Joe Flom who had set the strategy behind the Conoco fight. According to Gerard Colby Zilg, Shapiro was also a friend of Edgar Bronfman and had collaborated with him on various business deals.

A fierce fight ensued between Seagrams and Du Pont for Conoco's outstanding shares. Soon, Mobil Oil joined the takeover frenzy, as the prices for Conoco stock skyrocketed from $50 a share to almost $120. On top of cash, the Du Pont Company made what proved to be a fatal offer, 1.7 shares of Du Pont stock for each share of Conoco stock. In early August, the Justice Department approved the merger between Du Pont and Conoco. Tenders were recorded and tallied, and soon the figures were apparent: 37.9 million shares had been bought, and 9.4 million shares had been exhanged for 1.7 Du Pont shares each. Du Pont had paid $7.57 billion for an oil company that it didn't need and that was worth billions of dollars less than what was paid for it. But Du Pont had won, or so the company's leaders thought.

The trap was sprung. Edgar Bronfman was the winner, and the Du Ponts the losers. Seagrams had bought 28 million Conoco shares, or 32.6 percent. Turning in these shares to Du Pont, Seagrams collected 47.6 million Du Pont shares, or 20.2 percent of Du Pont's 236 million shares outstanding. Says Zilg (p. 863): "It came as quite a shock in Wilmington. 'I almost stopped drinking V.O. for a couple of weeks,' Governor [Pete] Du Pont later said with a smile." The Du Pont family owned only 19.8 percent of the shares, which meant that, in less than 24 hours, the Bronfmans had become the majority stockholders of the company.

The Du Pont board of directors was frantic. Bronfman could continue buying another 20 percent of Du Pont stocks, and throw them all out in the street. The Du Ponts had no defenses; following Shapiro's leadership, the company had used all its ready cash on Conoco. So the Du Ponts made a deal with the Bronfmans, promising the whiskey magnates that they could have a strong representation on the Du Pont board of directors if they limited their stock purchases to 25 percent of the outstanding shares. The final deal was very lucrative for the Bronfmans: Seagrams would not only earn dividends on the stock, but also 25 percent of the earnings of the Du Pont Company. The Bronfmans appointed Irving Shapiro to the board of directors of Seagrams.

In 1981, Edgar and Charles Bronfman were made members of

Du Pont's board of directors, but their influence on company decisions was small compared to that of the Du Pont family. By 1986, the Bronfmans had acquired a much greater voice and would soon assume near total control of the policies of the company.

This final step in the takeover is described in an interesting front-page article in the *The Wall Street Journal,* July 17, 1987. Reporter Laurie Hays writes:

> Today, Du Pont managers aren't likely to overlook the Bronfmans. Once thought of as outsiders who had sneaked into the company through the back door, the Canadian-born distillers now are playing a visible and vigorous role as advisers to the world's largest chemical concern. . . .
>
> Inside the company, the Bronfmans are referred to as "the new Du Ponts."
>
> On the board of directors, Seagrams appointees now outnumber Du Pont family members. They get special briefings that officers once worked up only for the Du Ponts. A team of Seagrams advisers monitors the investment, and Seagrams' board even meets once a year at Du Pont's Wilmington, Del., headquarters. . . .
>
> The Canadians have their say about Du Pont affairs not just at board meetings but also, says Du Pont Chairman Richard Heckert, in "the kind of dialogue that is half social, half business.". . .
>
> The Bronfmans "don't walk down the hall saying, 'You're hired and you're fired,'" explains Harold Fieldsteel, a former Seagrams executive vice president and retired Du Pont director. "But they are actively involved." On a scale of one to 10, he says, "on major questions, their influence is an eight, nine or 10."
>
> The Bronfmans will soon gain even more influence through a new retirement policy for Du Pont directors, adopted last fall but little noticed at the time. The policy calls for 10 Du Pont family members and retired executives . . . to leave the board by 1989. . . .
>
> The resulting shrinkage of the board not only will accommodate the 25 percent Seagram representation provided by a 1986 standstill agreement but also, in a break with tradition, reduce . . . the representation of family members

and current or former Du Pont executives. "We have to make some room for the new owners," says Irenee Du Pont, a director and retired official.

Adds Mr. Shapiro, who will leave the board next year, "This is a tacit recognition that the Bronfmans are the main players" [p. 1]

Not all directors of Du Pont were so happy. Hays reports the comments of one disgruntled director, "Selling booze and wine isn't the same thing as running a worldwide chemical company." And it is the booze that has been hurting the Bronfmans recently. According to the Dec. 18, 1989 cover story in *Business Week*:

Liquor consumption has been falling for a decade, and whiskey, Seagrams' traditional strength, has been hit hardest. Once the world's premier liquor company, Seagrams has been elbowed aside by two powerful and acquisitive companies, Grand Metropolitan PLC and Guiness PLC. Seagrams' profits have nearly tripled since 1981, to an expected $710 million this year on revenues of $6 billion. But that's misleading: About 75 percent of Seagrams' profits come not from liquor but from Du Pont Co., in which it acquired a 20 percent stake in 1980 [p. 54].

Indeed, Seagrams has stayed alive only by keeping 25 percent of Du Pont's earnings as its own.

So far, the ban on CFCs has proven quite profitable. The costs of manufacturing CFCs has remained the same, yet prices have risen to 10 times what they were in 1988. Replacement chemicals for the CFCs will be even more profitable. According to one expert, Edgar Bronfman personally stands to make more than $10 billion during the early 1990s, as a result of the ozone depletion swindle.

The Bronfmans are perfect corporate environmentalists. The new owners of the Du Pont Company have ties to the environmental movement that go back to nearly two decades before the CFCs issue. The unsavory nature of this relationship is illustrated by the fact that Charles Bronfman and his former wife, Barbara Bronfman, financed the activities of one Ira Einhorn, the leading figure of the 1970 Earth Day celebrations and a fugitive from the United States as the result of his 1977 indictment for the murder of his girlfriend Holly Maddux.*

The Resistance

It may seem that the picture painted in this chapter is very bleak. There is hope, however. On April 23, 1990 Lewis Du Pont Smith, heir to the positive tradition of the Du Pont family, made a major intervention into the company's annual shareholders meeting. Du

* Charles Bronfman's extensive connections to the New Age and environmental movements are documented in *The Unicorn's Secret,* a biography of environmentalist guru Ira Einhorn by Stephen Levy.

Levy reports that among Einhorn's influential friends: "were Canadians Charles and Barbara Bronfman, the former an heir of Samuel Bronfman, founder of the Seagrams liquor empire. . . . Barbara Bronfman had a strong interest in the psychic fields that Ira was plugged into, and not only did the Bronfmans become key names in Ira's network but Ira became a familiar visitor to their home in Montreal [p. 175]."

Who is Ira Einhorn and why would the Bronfmans become so close to him? For those who are old enough to remember, the main event of the 1970 Earth Day was a series of rallies, speeches, teach-ins, and demonstrations culminating in an environmental Woodstock on April 22, which was televised by CBS with Walter Cronkite as a host and Einhorn as the master of ceremonies. The finale of the televised show was a speech by Senator Edmund S. Muskie, who proposed a series of actions and legislative initiatives to bring environmentalism into mainstream American politics. One such legislative initiative was the 1970 Clean Air Act, a disaster that has cost the economy somewhere between $1.2 and $2.0 trillion, with little benefit to the environment or to people.

Levy reports: "Casual observers assumed that Earth Day was largely an Ira Einhorn show, and in future utterances, Einhorn fostered that impression himself. By 1979, he was claiming that 'I planned, directed, and emceed the entire event.'"

With this statement, Einhorn revealed, perhaps unintentionally, his importance to the emerging environmentalist movement. Earth Day was a watershed event, which helped to shift the attention of the American public and the energies of the antiwar and civil rights movements into a new movement: environmentalism. Clearly environmentalism was sponsored from the highest levels of the U.S. establishment: Earth Day was proposed by Senator Gaylord Nelson and funded by America's Eastern Establishment elites.

Einhorn was one of the leading gurus of the New Age. One of his roles in the organizing of Earth Day was to convince the college students and hippies that Earth Day was not a CIA operation. According to Levy, the behind-the-scenes organizers of Earth Day, like Edward Furia, used Einhorn. Levy says that Furia "needed Ira . . . [his] presence helped diffuse those in the radical community who were accusing Earth Week of draining the energy from the more righteous antiwar and civil rights movements. In one meeting between the [Earth Week] committee and some local interest groups, a black politician, Charles Bowser, accused the organizers of being 'co-opted' by the Nixon administration, and stormed out of the session. . . . Einhorn leapt to his feet. . . . Ira got on the phone and started calling gurus and famous people in the counterculture. All of whom he knew" (p. 118). Ira got commitments from Allen Ginsberg, Alan Watts, *Dune* author Frank Herbert, and Nobel Laureate George Wald.

Pont Smith published a pamphlet denouncing the Bronfmans and calling for a blue ribbon commission to investigate the CFCs issue. The pamphlet, titled "A Memorandum On the Horrifying Economic and Strategic Implications of the Company's Scientifically Flawed Policy with Respect to CFCs and 'Ozone Depletion,' " was handed

---

Einhorn, in many ways, was an ideal figure to have up front for Earth Day. At the same time he was a hippie guru, he also had extensive connections into the corporate world. From his base of operations in the Powelton Village section of Philadelphia, Einhorn had been made into a celebrity by the news media. He was close friends with such luminaries as Jerry Rubin and Abbie Hoffman and was a leading member of the Neo-American Church, founded by Timothy Leary, which used LSD as part of its religious sacrament.

In 1967, Einhorn, Curtis Kubiak, and Grant Schaefer, also leaders of the Neo-American Church, set up the University of Pennsylvania's Free University, and started openly teaching the use of LSD and other psychedelic drugs to college students. Einhorn's course, "Analogues to the LSD Experiment," had hundreds of students. The topics of this and other courses he taught ranged from expanding the mind through drug use, to air pollution and the dangers of population growth. It was not until 1969 that Einhorn decided upon a new focus, to launch "an international conspiracy to make the planet livable," which led to the 1970 Earth Day, and his subsequent activities.

In the 1970s, Einhorn cultivated a vast network of followers and sponsors, which eventually included heads of corporations, top scientists around the world, and believers in the paranormal, out-of-body experiences, ESP, UFOs, paganism, and other assorted New Age beliefs. Prominent among the believers were Charles and Barbara Bronfman.

Despite this high-level sponsorship, Einhorn turned himself from an asset into a liabilty in 1979.

On March 28, 1979, Philadelphia detective Michael Chitwood rang the doorbell of Einhorn's apartment in the Powelton section of the city. Chitwood showed Einhorn a 35-page search warrant and entered the apartment, followed by six other policemen.

Chitwood went straight for a closet in the back of Einhorn's apartment, journalist Martin Gardner reported in the *Skeptical Inquirer*. After breaking the lock and opening the closet, he pulled out a black steamer trunk. Chitwood broke the lock and opened the trunk. Inside, wrapped in plastic, was a mummified and partially decomposed human body. It had been cut to pieces, drained of blood, and packed in styrofoam chips. The front and sides of the skull had been fractured at a dozen places.

The year-and-a-half long search for Helen "Holly" Maddux, Einhorn's girlfriend, had ended.

Gardner wrote: " 'It looks like Holly's body,' said the detective. 'You found what you found,' Einhorn replied."

Newspapers in the trunk indicate that Einhorn had murdered Holly on Sept. 11, 1977, the night of her disappearance. Friends say that Holly had planned to end her relationship with Einhorn and move out of his apartment. Shortly after Sept. 11, Einhorn's neighbors began to complain about a foul odor emanating from the apartment. The odor became so horrid that the next summer the neighbors had to abandon their apartment for a while.

Einhorn's influential sponsors quickly came to his aid. Arlen Specter, who had served two terms as Philadelphia's district attorney, signed on as his lawyer. Within

out to 1,000 stockholders. Du Pont Smith took the floor during the meeting and denounced the Bronfmans and their policies as criminal.

In the pamphlet, Lewis Du Pont Smith states: "The future well-being of hundreds of millions of human beings and of the Du

---

months of Einhorn's trial, Specter would become a U.S. senator, a post he still holds. Previously, Specter's claim to fame was his participation in the Warren Commission that investigated the assassination of John F. Kennedy. Specter conceived and argued for the discredited "single-bullet" theory of Kennedy's assassination.

Specter did a fine job for Einhorn. On April 3, at the hearing to determine whether Einhorn would receive bail, the courtroom was packed with influentials—top corporate executives, members of Philadelphia's patrician elite, and other famous figures who had come to testify as Einhorn's character witnesses. The judge, evidently impressed, set a low $40,000 bail. The bond was posted by Barbara Bronfman.

If Einhorn did kill Holly Maddux—as all the evidence indicates he did—it is possible that she was not his only victim. Martin Gardner points out that Einhorn had tried to kill women several times before. In 1962, he almost strangled to death a young Bennington College student. Writing about the incident in his notebook, Einhorn said, "To kill what you love when you can't have it seemed so right. . . . insanity, thank goodness, is only temporary." In 1966, "Einhorn almost killed a Penn undergraduate by bashing her in the head with a Coke bottle," according to Levy.

In January 1981, a few weeks before his murder trial, Einhorn fled the United States and went into hiding in Great Britain. With him was his new girlfriend, Jeanne Marie Morrison, whom he had met, ironically, at Charles and Barbara Bronfman's estate in Charlottesville, Virginia.

After Einhorn fled, the first place the police looked for him was at Charles Bronfman's estate in Montreal. Einhorn had used Bronfman as his reference when he rented his Philadelphia apartment. The police called Bronfman, but neither Charles nor Barbara would come to the phone.

From Great Britain, Einhorn traveled to Ireland, where he rented a flat in Dublin under his own name. The landlord, Denis Weaire, lived in the same house, and got to know Einhorn quite well, to the point that Weaire left Einhorn in charge of the house in April 1981, when Weaire's family traveled to the United States. Once in Chicago, the Weaires discovered their tenant was wanted as a murderer. On April 16, Dennis Weaire went to the FBI office in Chicago and identified a picture of Einhorn as the person living in his house.

As Levy describes it, ". . . [O]ne would have expected the FBI to do something about a positive identification of a fugitive wanted for murder. . . . But when Dennis Weaire returned to Dublin on Sunday, April 19, no one was there to meet him, as the FBI office in Chicago had promised" (p. 329). Weaire then went to the Irish police, who refused to arrest Einhorn. Weaire was given no choice but to go back to his house in Dublin and throw Einhorn out.

Back in the United States, the FBI did not even bother to inform the Philadelphia Police Department that Einhorn had been found. It was not until several weeks later that the officer in charge of the case, Sergeant Richard King, heard rumors that Einhorn had been spotted in Ireland. Unlike the FBI, the Philadelphia police were very eager to arrest Einhorn, and on October 14, 1981, Richard King personally flew to Dublin. According to Levy, "King's first efforts were to contact Irish authori-

Pont Company is at stake. The issue is the full backing that the corporation is giving to ecological extremists and the banning of CFCs. . . . From a financial standpoint, it may seem to make a lot of sense. Du Pont will make tens of billions of dollars in profits over the next two decades from the chemicals it has patented to replace CFCs. From scientific and a moral standpoint, however, the Corporation's policy is dead wrong."

Du Pont Smith warned: "The banning of CFCs will result in the collapse of the worldwide food refrigeration chain, and experts estimate that as many as 20 to 40 million people will die of hunger,

---

ties to determine what could be done if he actually *found* Ira Einhorn. 'This was an inquiry that no one seemed to want to answer,' he noted in his report. After being shuttled around in bureaucracy, King finally got the answer—Ira Einhorn was not to be arrested. . . . King was told specifically, 'Do not molest Einhorn—he is not to be touched' " (p. 330). Einhorn, meanwhile, had moved to the richest section of Dublin, renting a luxury apartment four blocks away from the U.S. embassy.

The next break in the case came in 1983, when Barbara Bronfman contacted Ian Summers, a book packager with whom Einhorn had worked before he fled the United States. In a luncheon meeting, Mrs. Bronfman requested that Summers sell a manuscript she had. The manuscript had a secret author, and she refused to either confirm or deny that it was Einhorn. Levy reports that Summers was very upset:

" 'How peculiar that someone of your status and your husband's status would have befriended this man,' he recalls saying to her. 'What was in it for you? What did you like about Ira that made you really go this far?'

"And she said, 'Ira was the best friend that anybody could ever have [p. 333].' "

After reading the manuscript, it was evident to Summers that Einhorn had written it. Mrs. Bronfman, however, never called him back.

Late in 1986, Einhorn's girlfriend, Jeanne Marie Morrison, returned to the United States, and contacted her family. She had left Einhorn in 1984, but had continued to live for two years as a fugitive. In interviews with the police, Morrison gave details of her life with Einhorn, but none that would lead to his capture. But she did provide the police with an answer to one of their most important questions. Who provided the money for Einhorn to live high on the hog in Dublin? Morrison gave the police one name: Barbara Bronfman.

In 1986, Einhorn was discovered again in Dublin living under a pseudonym. Immediately, he fled his luxurious apartment and has not been heard from since. According to Morrison, he had spent his time in Ireland on a thorough study of pagan history, the druids, and the Roman empire, and was in close contact with many members of the New Age/environmentalist network. Among these were New Age gurus William Irwin Thompson and Hazel Henderson, whose activities are detailed in Chapter 11.

Charles and Barbara Bronfman divorced in 1982, and there is no evidence that Charles has continued to financially and politically support Einhorn since then. Nevertheless, it is clear that the new owner of the Du Pont Company has had personal involvement with the leadership of the New Age environmentalist movement.

Phil Valenti

Lewis Du Pont Smith at a press conference Aug. 23, 1990 in front of Du Pont Corporate Headquarters in Wilmington, Delaware, where he called for the resignation of Edgar Bronfman from the board of directors.

starvation, and food-borne diseases every year as a result of this. All for what?

"The stockholders must wake up to the fact that the Board of Directors, without the stockholders' approval, or consultation, has embarked on a course that may destroy the corporation at the point that the ozone depletion issue is exposed as a fraud in the news media."

Du Pont Smith ended his statement with the following appeal:

I request this proxy statement be brought to the floor for a vote, and request your help in returning the Du Pont Corporation to its glorious role of a company in the service of nation-building. We must use the company's technology and know how to feed and clothe this desperate world. Political prisoner Lyndon LaRouche, who has outlined the economic policies required to end this depression, must be freed, and we must set to work on the kind of great infrastructural development projects that are required.

239

His fight for a policy change at Du Pont Company continues today.

References

Richard Elliot Benedick, 1991. *Ozone Diplomacy, New Directions in Safeguarding the Planet*. Cambridge, Mass.: Harvard University Press.

Editors of *Executive Intelligence Review*, 1986. *Dope, Inc.: Boston Bankers and Soviet Commissars*. New York: New Benjamin Franklin House Publishing Company.

William Engdahl, 1989. "Du Pont, ICI Behind the 'Ozone' Scare?" *Executive Intelligence Review* (May 12), Vol. 16, No. 20, pp. 14–15.

Martin Gardner, 1989. "The Unicorn at Large," *Skeptical Inquirer* (Fall), pp. 16–20.

Laurie Hays, 1987. "Bronfmans of Seagrams Take Increasing Role in Du Pont Co. Affairs. They Press Firm to Update Its Lines and Its Methods, While Profiting Nicely," *The Wall Street Journal* (July 17), p. 1.

Richard L. Hudson, 1990. "Giant Chemical Companies Should Prosper from Expected 'Ozone Friendly Accord,' " *The Wall Street Journal Europe* (June 29–30), p. 7.

David Kirkpatrick, 1990. "The Environment: Business Joins the New Crusade," *Fortune* (Feb. 12), pp. 44–50.

Stephen Levy, 1988. *The Unicorn's Secret. Murder in the Age of Aquarius*. Englewood Cliffs, N.J.: Prentice Hall Press.

Peter C. Newman, 1979. *King of the Castle: The Making of a Dynasty: Seagram's and the Bronfman Empire*. New York: Atheneum.

Andrea Rothman, 1989. "The Maverick Boss at Seagrams: Edgar Bronfman, Jr. Is Reshaping the Company for a Tougher World," *Business Week* (Dec. 18), pp. 54–59.

Lewis Du Pont Smith, 1991. "A Memorandum on the Horrifying Economic and Strategic Implications of the [Du Pont] Company's Scientifically Flawed Policy with Respect to CFCs and 'Ozone Depletion,' " (April 24).

Maury Terry, 1987. *The Ultimate Evil, An Investigation into America's Most Dangerous Satanic Cult*. New York: Doubleday.

Gerard Colby Zilg, 1974. *Du Pont: Behind the Nylon Curtain*. Englewood Cliffs, N.J.: Prentice Hall. (The second edition, 1984, published by Lyle Stuart (Seacaucus, N.J.), contains the Bronfman material.)

10

Who Owns the Environmental Movement?

The close collaboration of the environmental movement with the giant chemical companies in their drive to ban CFCs poses the fundamental issue that will be raised in this chapter. It will be a great shock to many readers, but the fact is that the environmental movement was created and is to this day financed and directed by the leading aristocratic families of the United States and Europe, especially Great Britain.

This is one of the most controversial assertions of this book, but ironically, it is one of the easiest to prove. All one has to do is trace the origin of the money that finances the environmental movement. The more difficult concept to convey is the nature of environmentalism itself. The basic problem is that environmentalism is not a rigorous definition of anything. Recent polls indicate that more than 90 percent of the American public consider themselves "environmentalists." Yet every environmental proposal submitted to popular vote over the past four years has been soundly defeated.

Why the contradiction? The answer lies in the questions asked by the polls—questions that are not usually reported in the press. For example, when asked if they like trees, open spaces, mountains, and wildlife and are opposed to toxic chemical waste dumps, more than 90 percent of those polled answer affirmatively. Why not? Only a deranged person would be opposed to the beauty of nature. Only those who actually enjoy those wastelands known as parking lots, shopping centers, malls, and their equally ugly modern architecture would answer in the negative.

However, when those being polled are asked if they believe that man is the source of all evil and if, to preserve nature, they support forced abortion and sterilization to control world population along with the elimination of automobiles, electricity, hot showers, flushing toilets, refrigeration, beef, fertilizers, pesticides, and insecticides, the answer is a solid kick in the rear. Therein lies the inherent contradiction in the term "environmentalism."

This problem is exacerbated by the fact that the majority of the members of mainstream environmental groups supports the first set of beliefs but is decidedly against the second set of beliefs. At the same time, the leaders of the majority of the environmental groups are fanatical advocates of the second set of beliefs (as long as it is not they who have to give up their cars). The fact is that most environmentalists are moral and loving human beings who would be shocked to learn what they are really supporting.

We intend to expose the nature of this deceit. Lacking any other term, we mean by environmentalism the set of beliefs that considers that people are the problem, that animals and insects are more important than human beings, that modern science and technology are evil, and that the world's population—especially darker-skinned people—must be dramatically reduced to what are euphemistically called "sustainable" levels.

Twenty-five years ago, those who believed that Mother Nature comes first and humankind second were part of an insignificant fringe, considered radical by most Americans. These environmentalists were visible mostly at the level of the antinuclear street demonstration, where marijuana smoke wafted around "Back to Nature" posters on display. Today, however, what used to be extremist "environmentalist" ideology has become mainstream, permeating American institutions at every level, from corporate boardrooms to the Federal Reserve, the Congress, the White House, the churches, homes, and schools.

Official lore from the environmental movement's publications asserts that the movement emerged from the grass roots. The truth, however, is that funding and policy lines come from the most prestigious institutions of the Eastern Liberal Establishment, centered around the New York Council on Foreign Relations, and including the Trilateral Commission, the Aspen Institute, and a host of private family foundations. These foundations and institutes are the policy-making and implementation arms of what is known as the Eastern Establishment. They are run largely by leading members of the Anglo-American blueblood elites. Over the past 25 years, these foundations have poured hundreds of millions of dollars every year into the anti-industry, environmentalist, and population-control campaigns of hundreds of "public interest" groups. Additional billions have been poured into sponsoring university departments of "environmental studies," which now turn out thousands of professional environmentalists every year. Many of these professional environmentalists act on the basis of political ideology, not hard science.

This network of foundations created environmentalism, moving it from a radical fringe movement into a mass movement to support the institutionalization of antiscience, no-growth policies at all levels of government and public life. As prescribed in the Council on Foreign Relations *1980s Project* book series, environmentalism has been used against America's economy, against such targets as high-technology agriculture and the nuclear power industry. This movement is fundamentally a green pagan religion in its outlook. Unless defeated, it will destroy not only the economy, but also the Judeo-Christian culture of the United States, and has in fact come perilously close to accomplishing this objective already.

The vast wealth of the environmentalist groups may come as a shock to most readers who believe that these groups are made up of "public interest," "nonprofit" organizations that are making great sacrifices to save the Earth from a looming doomsday caused by man's activities. In fact, the environmental movement is one of the most powerful and lucrative businesses in the world today.

Funding from the Foundations

There are several thousand groups in the United States today involved in "saving the Earth." Although all share a common philosophy, these groups are of four general types: those concerned, respectively, with environmental problems, population control, animal rights, and land trusts. Most of these groups are very secretive about their finances, but there is enough evidence on the public record to determine what they are up to.

Table 10.1
ANNUAL REVENUES OF A FEW ENVIRONMENTAL GROUPS
(U.S. dollars, 1990,1991)

Organization	Revenues
African Wildlife Foundation	4,676,000
American Humane Association	3,000,000
Center for Marine Conservation	3,600,000
Clean Water Action	9,000,000
Conservation International	8,288,216
The Cousteau Society	14,576,328
Defenders of Wildlife	6,454,240
Earth Island Institute	1,300,000
Environmental Defense Fund	16,900,000
Greenpeace International	100,000,000
Humane Society	19,237,791
Inform	1,500,000
International Fund for Animal Welfare	4,916,491
National Arbor Day Foundation	14,700,000
National Audubon Society	37,000,000
National Parks Conservation Association	8,717,104
National Wildlife Federation	77,180,104
Natural Resources Defense Council	16,000,000
Nature Conservancy	254,2511,717
North Shore Animal League	26,125,383
Population Crisis Committee	4,000,000
Rails-to-Trails Conservancy	1,544,293
Sierra Club	40,659,100
Student Conservation Association, Inc.	3,800,000
Trust for Public Land	23,516,506
Wilderness Society	17.903,091
Wildlife Conservation International	4,500,000
WWF/Conservation Foundation	51,555,823
Zero Population Growth	1,300,000
Total	**$830,367,693**

Sources: *Buzzworm*, Sept.–Oct. 1991; *The Chronicle of Philanthropy*, March 23, 1992.

Table 10.1 lists the annual revenues of a sampling of 30 environmental groups. These few groups alone had revenues of more than $830 million in 1990. This list, it must be emphasized, by no means includes all of these envirobusinesses. It is estimated that there are more than 3,000 so-called nonprofit environmental groups in the United States today, and most of them take in more than a million dollars a year. The Global Tomorrow Coalition, for example, is made up of 110 environmental and population-control groups, few of which have revenues less than $3 million per year. The Nature Conservancy, with revenues of $254 million per year and land holdings of more than 6 million acres worth billions of dollars, is just the best known of more than 900 land trusts now operating in the United States.

Table 10.2, lists the grants of 35 foundations to two heavily funded and powerful environmentalist groups—the Environmental Defense Fund and the Natural Resources Defense Council— for the year 1988.

The data available from public sources show that the total revenues of the environmentalist movement are more than $8.5 billion per year. If the revenues of law firms involved in environmental litigation and of university environmental programs were added on, this figure would easily double to more than $16 billion a year. This point is emphasized in Table 10.3, which lists the 15 environmental groups receiving grants for environmental lawsuits and protection and education programs.

To get an idea of how much money this is, the reader should consider that this income is larger than the Gross National Product (GNP) of 56 underdeveloped nations (Table 10.4). The 48 nations for which the latest GNP figures were available have a total population of more than 360 million human beings. Ethiopia, for example, with a population of 47.4 million human beings, many starving, has a GNP of only $5.7 billion per year. Chad, with 5.4 million inhabitants, has a GNP that is barely higher than the revenues of those groups listed in Table 10.1. Not a single nation in Central America or the Caribbean has a GNP greater than the revenues of the U.S. environmental movement.

With these massive resources under its control, it is no surprise that the environmentalist movement has been able to set the national policy agenda. There is no trade association in the world with the financial resources and power to match the vast resources of the environmental lobby. In addition, it has the support of most of the news media. Opposing views and scientific refutations of environmental scares are most often simply blacked out.

Where do the environmental groups get their money? Dues from members represent an average of 50 percent of the income of most groups; most of the rest of the income comes from foundation grants, corporate contributions, and U.S. government funds. Almost every one of today's land-trust, environmental, animal-rights, and population-control groups was created with grants from one of the elite foundations, like the Ford Foundation and the Rockefeller

Table 10.2
WHO OWNS THE ENVIRONMENTAL MOVEMENT?
FOUNDATION GRANTS TO EDF AND NRDC
(U.S. dollars, 1988)

Foundation	EDF	NRDC
Beinecke Foundation, Inc.		850,000
Carnegie Corporation of New York	25,000	
Clark Foundation		150,000
Columbia Foundation		30,000
Cox Charitable Trust		38,000
Diamond Foundation		50,000
Dodge Foundation, Geraldine	75,000	10,000
Educational Foundation of America	30,000	75,000
Ford Foundation	500,000	
Gerbode Foundation	50,000	40,000
Gund Foundation	85,000	40,000
Harder Foundation	200,000	
Joyce Foundation	75,000	30,000
MacArthur Foundation		600,000
Mertz-Gilmore Foundation	75,000	80,000
Milbank Memorial Fund		50,000
Morgan Guaranty Charitable Trust	5,000	6,000
Mott Foundation, Charles Stewart	150,000	40,000
New Hope Foundation, Inc.		45,000
New York Community Trust	35,000	
Noble Foundation, Inc.	20,000	35,000
Northwest Area Foundation		100,000
Packard Foundation	50,000	37,000
Prospect Hill Foundation		45,000
Public Welfare Foundation	150,000	
Robert Sterling Clark Foundation	50,000	40,000
Rockefeller Brothers Fund		75,000
San Francisco Foundation		50,000
Scherman Foundation	40,000	50,000
Schumann Foundation		50,000
Steele-Reese Foundation		100,000
Victoria Foundation, Inc.	35,000	35,000
Virginia Environmental Endowment		25,000
W. Alton Jones Foundation	100,000	165,000
Wallace Genetic Foundation, Inc.	80,000	65,000
William Bingham Foundation	1,000,000	150,000
Total*	**2,885,000**	**3,236,000**

* The total includes some smaller foundation grants not listed here.

Source: *The Foundation Grants Index*, 1989, 1990.

Table 10.3
TOP 15 RECIPIENTS IN ENVIRONMENTAL LAW,
PROTECTION, AND EDUCATION
(By single highest grant amount, U.S. dollars, 1987)

Recipient	Foundation	Grant in $
1. World Resources Institute	MacArthur Foundation	15,000,000
2. World Resources Institute	MacArthur Foundation	10,000,000
3. Nature Conservancy	R.K. Mellon Foundation	4,050,000
4. Nature Conservancy	Champlin Foundations	2,000,000
5. Oregon Coast Aquarium	Fred Meyer Charitable Trust	1,500,000
6. International Irrigation Management Institute	Ford Foundation	1,500,000
7. Open Space Institute	R.K. Mellon Foundation	1,400,000
8. International Irrigation Management Institute	Rockefeller Foundation	1,200,000
9. Chicago Zoological Society	MacArthur Foundation	1,000,000
10. Native American Rights Fund	Ford Foundation	1,000,000
11. Wilderness Society	R.K. Mellon Foundation	800,000
12. World Resources Institute	A.W. Mellon Foundation	800,000
13. University of Arkansas	W.K. Kellogg Foundation	764,060
14. National Park Service	Pillsbury Co. Foundation	750,000
15. National Audubon Society	A.W. Mellon Foundation	750,000

Source: *Environmental Grant Association Directory,* 1989.

Foundation. These "seed grants" enable the radical groups to become established and start their own fundraising operations. These grants are also a seal-of-approval for the other foundations.

The foundations also provide funding for special projects. For example, the Worldwatch Institute received $825,000 in foundation grants in 1988. Almost all of that money was earmarked specifically for the launching of a magazine, *World Watch,* which has become influential among policy-makers, promoting the group's antiscience and antipopulation views. The Worldwatch Institute's brochures report that it was created by the Rockefeller Brothers Fund to "alert policy makers and the general public to emerging global trends in the availability and management of resources—both human and natural."

Foundation grants in the range of $20 to $50 million for the environmental cause are no longer a novelty. In July 1990, the Rockefeller Foundation announced a $50 million global environmental program. The specific purpose of the program is to create an elite group of individuals in each country whose role is to implement and enforce the international environmental treaties now being negotiated. Kathleen Teltsch reported in *The New York Times* (July 24, 1990): "As an initial step, the five-year program will

Table 10. 4
UNDERDEVELOPED NATIONS WHOSE GROSS NATIONAL PRODUCT (GNP) IS LESS THAN THE ANNUAL REVENUES OF U.S. ENVIRONMENTAL GROUPS (1990)

Country	GNP (billions $)	Population (millions)
Bhutan	0.25	1.4
Laos	0.70	3.9
Lesotho	0.71	1.7
Chad	0.86	5.4
Mauritania	0.91	1.9
Somalia	1.00	5.9
Yemen	1.03	2.4
Central African Republic	1.10	2.9
Botswana	1.21	1.2
Burundi	1.22	5.1
Togo	1.26	3.4
Malawi	1.36	8.0
Mozambique	1.49	14.9
Benin	1.72	4.4
Burkina Faso	1.79	8.5
Mali	1.84	8.0
Congo	1.91	2.1
Madagascar	1.96	10.9
Mauritius	1.98	1.1
Rwanda	2.14	6.7
Niger	2.19	7.3
Zambia	2.20	7.6
Guinea	2.32	5.4
Haiti	2.39	6.3
Jamaica	2.57	2.4
Papua New Guinea	3.00	3.7
Nepal	3.24	18.0
Gabon	3.27	1.1
Bolivia	3.93	6.9
Tanzania	3.95	24.7
Trinidad and Tobago	4.02	1.2
Honduras	4.13	4.8
Uganda	4.54	16.2
Senegal	4.55	7.0
Costa Rica	4.56	2.7
El Salvador	4.70	5.0
Paraguay	4.72	4.0
Panama	4.88	2.3
Dominican Republic	4.97	6.9
Ghana	5.60	14.0
Ethiopia	5.69	47.4
Jordan	5.85	3.9
Sri Lanka	6.97	16.6
Oman	7.00	1.4
Uruguay	7.66	3.1
Guatemala	7.83	8.7
Kenya	8.29	22.4
Ivory Coast	8.62	11.2
Total		**362.0**

Figures were not available for Afghanistan, Kampuchea, Liberia, Sierra Leone, Angola, Lebanon, Nicaragua, and Vietnam

Source: *World Development Report 1990: Poverty,* The World Bank (New York, London: Oxford University Press, 1990)

assist hundreds of young scientists and policy makers in developing countries to create a worldwide network of trained environmental leaders, who will meet regularly at workshops, sharing information and discussing strategy. Through the international network, the foundation wants to encourage efforts to build environmental protection into governments' long-range economic planning. Other major elements would promote the drafting of international treaties to deal with forest, land, and water preservation, and hazardous waste disposal."

The foundations are run by America's top patrician families. These families channel billions of dollars into the organizations and causes they wish to support every year, and thereby exert enormous political clout. By deciding who and what gets funded, they determine the political issues up front in Washington, which are then voted on by Congress. It is all tax free, since the foundations are tax-exempt. The boards of directors of the large foundations are made up of some of the most powerful individuals in this country, and they always overlap with power brokers in government and industry.

One such individual was Thornton F. Bradshaw, who, until his recent death, was chairman and program director of the MacArthur Foundation and a trustee of the Rockefeller Brothers Fund and the Conservation Foundation. At the same time, Bradshaw was chairman of the RCA Corporation. and a director of NBC, the Atlantic Richfield Corp., Champion International, and First Boston, Inc. Bradshaw was also a member of the Malthusian Club of Rome and director of the Aspen Institute of Humanistic Studies, organizations that have played a critical role in spreading the "limits to growth" ideology of the environmental movement.

Another individual perhaps better known to readers is Henry A. Kissinger, former U.S. secretary of state and a trustee of the Rockefeller Brothers Fund. For years Kissinger was the director of the fund's Special Studies Project, which was in charge of special operations.

Corporate Contributions

Another huge source of contributions to the environmental movement is private corporations. Unlike tax-exempt foundations, however, corporations are not required by law to report what they do with their money, so it is difficult for an independent researcher to estimate the level of funding for the environmentalist movement from business and industry. There are watchdog groups, however, that have investigated these money flows and come up with startlingly large figures.

For example, the April 1991 newsletter of the Capital Research Center in Washington, D.C., which monitors trends in corporate giving, scathingly denounces those corporations it has discovered financing the environmentalists. The newsletter states that oil companies "are heavy financial supporters of the very advocacy groups which oppose activities essential to their ability to meet consumer needs." Further, it reports, "The Nature Conservancy's 1990 report reflects contributions of over $1,000,000 from Amoco, over $135,000 from Arco, over $100,000 from BP Exploration and BP Oil, more than $3,700,000 (in real estate) from Chevron, over $10,000 from Conoco and Phillips Petroleum and over $260,000 from Exxon."

From the scant information publicly available (largely annual reports from the major environmental groups), one can conservatively estimate that corporations contribute more than $200 million a year to the environmentalist movement.

This should come as no surprise. Over the past 20 years, giant corporations have discovered that by using environmental regulations they can bankrupt their competition, the small- and medium-sized firms that are the most active and technologically innovative part of the U.S. economy.

Compliance with environmental regulations is also big business. According to official figures from the federal government's Environmental Protection Agency (EPA), it costs the U.S. economy $131 billion today to comply with environmental regulations. That figure will have risen to more than $300 billion a year by the year 2000. The expenditures are a net drain on the economy, but while the nation is bankrupted, someone is profiting from the services and equipment sold. A look at classified advertisements in the papers today reveals that companies involved in environmental compliance are growing fast. Many of these corporations are contributing to the environmental movement.

Funds from the U.S. Government

There is a third area of funding for the environmental movement: the U.S. government itself. As reported in detail by Peter

Metzger, former science editor of the *Rocky Mountain News,* there are now thousands of professional environmentalists ensconced in the U.S. government. These environmentalists channel hundreds of millions of dollars in grants and favors to environmentalists and environmental groups under all kinds of guises. In a 1991 newspaper series, columnist Warren Brookes exposed how the federal Bureau of Land Management used the Nature Conservancy as a land broker, giving the antigrowth organization handsome profits.

The EPA doles out huge amounts of money to environmental groups to conduct "studies" of the impact of global warming and ozone depletion. President Bush has made the Global Climate Change program a priority, so while the Space Station, vaccinations for children, and other crucial projects have been virtually eliminated from the budget, $1.3 billion is available for studies of how man is fouling the Earth. Similarly, scientists who challenge global warming and ozone depletion as hoaxes do not receive a penny in funding, while those who scream doomsday receive tens of millions in research grants from the "climate change" program.

How much funding do the environmentalists receive from the federal government? Officially, the U.S. government gives away more than $3 billion a year in grants to support environmental groups and projects. The actual total, however, is impossible to estimate. A top-ranking official of the Department of Energy who spent two years attempting to cut off tens of millions of dollars in "pork-barrel" grants going to environmentalist groups, discovered that for each grant she was eliminating, environmentalist moles in the department added several new ones. The official resigned in disgust.

The environmentalist capture of Washington, which was consolidated during the Carter administration, produced radical changes in the Washington, D.C. establishment. This process of subversion was described by Metzger in a speech given in 1980, titled "Government-Funded Activism: Hiding Behind the Public Interest."

> For the first time in history, a presidential administration is funding a political movement dedicated to destroying many of the institutions and principles of American society. Activist organizations, created, trained, and funded at taxpayers' expense, and claiming to represent the public interest, are attacking our economic system and advocating its replacement

251

by a new form of government. Not only is this being done by means already adjudicated as being unconstitutional, but it is being done without the consent of Congress, the knowledge of the public, or the attention of the press.

It all began when President Carter hired individuals prominently identified with the protest or adversary culture . . . the appointment [by the Carter administration] of several hundred leading activists to key regulatory and policy-making positions in Washington resulted in their use of the federal regulatory bureaucracy in order to achieve their personal and ideological goals. Already accomplished is the virtual paralysis of new federal coal leasing, conventional electric generating plant licensing in many areas, federal minerals land leasing and water development, industrial exporting without complex environmental hearings, and the halting of new nuclear power plant construction. . . .

The consequences of those sub-cabinet appointees having then made their own appointments, and those having then made theirs, so that now, there are thousands of [environmentalist] representatives in government. . . .

According to Metzger, this new class, "enshrined in the universities, the news media, and especially the federal bureaucracy, has become one of the most powerful of the special interests."

Two Case Studies

Let us consider two case studies of how foundation-funded environmentalist organizations have virtually taken over national policy.

The Washington, D.C.-based Environmental Defense Fund (EDF) was created in 1969. The cover story is that it sprang from America's grass roots, after a group of Long Island citizens began having coffee clatches to discuss the threat of toxic chemicals. The truth is that EDF was created by grants from the leading Eastern Establishment foundations and these foundations have continued to support it. The Ford Foundation gave EDF its seed money in 1969. In 1988, EDF received $500,000 from the Ford Foundation, $1,000,000 from the William Bingham Foundation, $75,000 from the Joyce Foundation, $150,000 from the Mott Foundation, and $25,000 from the Carnegie Foundation, among others. Today, EDF has seven offices nationwide, more than 150,000 members, and an annual operating budget of $17 million.

The EDF made its name in the fight to ban DDT, which it accomplished with the help of Natural Resources Defense Council litigation in 1972—and with the cooperation of the EPA's administrator, William Ruckelshaus. Ruckelshaus ignored the scientific evidence presented at seven months of EPA hearings on DDT, and he ignored the decision of the EPA's hearing examiner *not* to ban DDT; instead, for what he admitted were political reasons, he banned this life-saving insecticide that was turning the tide on malaria. Thus "public perception" became established as more important than scientific evidence in environmental decisions.

In 1986, EDF helped to draft California's first sweeping environmental regulations in the form of the ballot initiative known as Proposition 65, which restricted the use of dozens of chemicals in industry and agriculture and has cost the California economy billions. EDF's goals for the 1990s include: defending against the so-called greenhouse effect; saving sea turtles and porpoises by shutting down the fishing industry; banning CFCs worldwide by the year 2000; saving the world's rain forests; passing legislation to prevent so-called acid rain; setting aside Antarctica as a permanent wildlife reserve; extending the chemical bans in California's Proposition 65 to the entire nation; and recycling all household and industrial waste material.

The Natural Resources Defense Council (NRDC), one of several of the legal arms of the environmentalist movement, was founded in 1970 with a massive infusion of funds from the Ford Foundation. Together with the Legal Defense Fund of the Sierra Club and the National Audubon Society, the NRDC took to the courts, filing dozens of lawsuits to block dams, shut down nuclear power-plant construction, and derail highway development projects. The NRDC and its cohorts also targeted federal regulators in the Environmental Protection Agency and other offices, forcing tightened controls on pollution and demanding the enforcement of statutory rules for clean air and rivers. The Clean Air Act of 1970 was a first fruit of these efforts.

Who funds these multi-million-dollar court battles? In 1988, the NRDC received grants of $75,000 from the Educational Foundation of America, $600,000 from the MacArthur Foundation, $165,000 from the W. Alton Jones Foundation, and $850,000 from the Beinecke Foundation, to name only a few.

A good chunk of this money ends up in the expense accounts and salaries of the Eastern Establishment bigwigs who run the environmentalist advocacy groups—or in the pockets of their lawyers. A 1990 cover story in *Forbes* magazine reports that the organizational network of consumer and environmentalist activist Ralph Nader is worth close to $10 million and receives ardent support in its anti-industry lawsuits from a circle of plaintiff attorneys with multi-million-dollar annual incomes (see Brimelow and Spencer 1990). Nader himself lives very well off the publicity stirred up from court cases. "Oh, God, limousines and nothing but the best hotels," *Forbes* quotes a former state Trial Lawyers Association official. "We got quite a bill when he [Nader] was in town." Nader lives in a $1.5 million townhouse in Washington, D.C. (owned by his sister) and commands up to five-figure fees each for between 50 and 100 speaking appearances per year.

Other environmentalist organization leaders also maintain an expensive lifestyle. In August 1983, reporter Nancy Shute gave a colorful description of the environmentalists-turned-establishment who had taken over Washington. Under the headline "Bambi Goes to Washington," Shute writes in *National Review:*

On December 1, 1982, barely two years after Ronald Reagan's election, hundreds of Washington lawyers and lobbyists munched pears and cheese and sipped Bloody Marys under the sparkling crystal chandeliers at the Organization of the American States headquarters, just two blocks from the White House. The conversation turned to politics, as do all Washington cocktail-party conversations. But the women in pearls and men in dark suits who shouted to be heard over the seven-piece dance band represented not Exxon or U.S. Steel or General Motors, but the nation's environmental lobby, celebrating the tenth birthday of the Environmental Policy Center, an influential Washington lobbying group and research institute. In the 13 years since Earth Day, the environmental presence in the capital has grown from a ragtag band dedicated to saving trees and whales to a formidable Washington institution.

Much of the environmental windfall has been spent on sleek new offices, on high-profile lobbyists like former Senator Gaylord Nelson and Carter Adminstration Interior Secretary Cecil Andrus . . . on high-priced economists and lawyers, and on millions of direct-mail pleas for more cash. . .[p. 924].

These environmentalists are unabashed about their affluence. Their conferences have become notorious for their plush locales (Switzerland, Beverly Hills, Sundance, and Aspen, for example).

The Campaign against CFCs

Both the EDF and NRDC played a leading role in the propaganda and legal campaign to ban CFCs.

In June 1974, Rowland and Molina's doomsday paper claiming CFCs would deplete the ozone layer was published in *Nature*. At that moment, however, the hottest topic in the news media was that chlorine emissions from the Space Shuttle would wipe out the ozone layer. It was not until September 1974 that articles on the CFCs threat started to appear.

In November 1974, the Natural Resources Defense Council joined the ozone debate, calling for an immediate ban on CFCs. In June 1975, the NRDC sued the Consumer Products Safety Commission for a ban on CFCs used in aerosol spray cans. The lawsuit was rejected by the commission in July 1975, on grounds that there was insufficient evidence that CFCs harm the atmosphere.

At that point, EPA administrator Russell E. Train intervened on behalf of the NRDC and proponents of the ozone-depletion theory, calling for all nations to cooperate in establishing worldwide guidelines on CFCs to avoid environmental disaster. Today Russell E. Train is head of the World Wildlife Fund/Conservation Foundation, a trustee of the Rockefeller Brothers Fund, and a top-ranking member of both the Trilateral Commission and the New York Council on Foreign Relations.

For the next two years, debate raged on the future of CFCs, with the NRDC, lavishly funded by the Ford and Rockefeller Foundations, playing a major role. While President Ford's top science advisers said the evidence was still not strong enough for an immediate ban on CFCs, other members of the administration moved to implement such a ban. One of them was Russell W. Peterson, chairman of the White House Council on Environmental Quality, who worked for a ban on the use of CFCs in aerosol cans as a first step toward the total banning of CFCs. Peterson made it clear that it did not matter that there was no scientific evidence against CFCs. According to Sharon Roan in *Ozone Crisis,* Peterson told the press: "I believe firmly that we cannot afford to give chemicals the same constitutional rights that we enjoy under the law. Chemicals are not innocent until proven guilty" (p. 83). Peterson today is the head of the National Audubon Society.

In October 1978, CFCs used as propellants in aerosol cans were banned in the United States.

The CFCs issue lay dormant for the next several years, until November 1984, when the NRDC started a new phase of the assault on CFCs with a suit against the EPA. The suit sought to force the EPA to place a cap on overall CFC production, as mandated under the EPA's Phase Two proposals. The NRDC argued that under the Clean Air Act, the EPA was required to regulate CFCs if they were deemed harmful to the environment. The group claimed the EPA had acknowledged this in its 1980 proposed regulations, which had not been implemented during the first four years of the Reagan administration.

As the NRDC relaunched its campaign against CFCs, a major political change was taking place in Washington, D.C. The leading proponents of technology, the space program, and economic development in the Reagan administration had been ousted by a series of media-orchestrated scandals—Interior Secretary James Watt, NASA Administrator James Beggs, and EPA Chief Anne Burford. Burford was replaced by the multimillionaire corporate environmentalist William Ruckelshaus, his second term as EPA administrator.

There was still no credible scientific evidence against CFCs; supposedly this changed in May 1985 with the publication of Joseph Farman's doomsday ozone-hole paper in *Nature* magazine. This article enabled the environmental lobby to start creating hysteria about CFCs once more, which set the wheels into motion that led to the signing of the Montreal Protocol in 1987.

In September 1986, the Du Pont Corporation announced its support for the banning of CFCs. By summer 1987, the environmental onslaught against CFCs was in full gear under the leadership of the well-funded NRDC. It was at that moment that the World Resources Institute received a $25 million grant from the MacArthur Foundation. According to Sharon Roan's book, *Ozone Crisis* (p. 204): "Economist Daniel J. Dudek of the Environmental Defense Fund provided a study on the cost of reducing ozone depletion. . . . At the World Resources Institute and Worldwatch

Institute, studies were completed to alert Americans to the effects of various ozone control policies. The Environmental Defense Fund, Friends of the Earth, and Sierra Club initiated public education campaigns and began pressuring industry to own up to its responsibility."

In September 1987, the Montreal Protocol was signed, calling for a 50 percent ban on CFCs by the year 2000.

The First Earth Day

At the same time that the environmental organizations were becoming a well-funded big business, their propaganda output was used to create popular support for the environmentalist cause in the United States. A turning point in the transformation of the environmentalist fringe into a radicalized mass movement was Earth Day 1970.

On April 22, 1970, thousands of college students and curious onlookers turned out to participate in the widely publicized Earth Day festivities in dozens of major U.S. cities. Folk music, antinuclear slogans, "Love Your Mother Planet Earth" posters, and college students were everywhere. On the surface it appeared to most observers that the nationwide rallies represented a grass roots movement to protest "the destruction of the environment." Nothing could be further from the truth. The Earth Day publicity stunt was part of a highly coordinated effort to create a climate of sympathy for Malthusian zero growth, where none yet existed in the United States. Earth Day was partly bankrolled by a $200,000 personal grant from Robert O. Anderson, at the time the president of Atlantic Richfield Oil Corporation, the president of the Aspen Institute for Humanistic Studies, and a personal protégé of University of Chicago zero-growth ideologue Robert Maynard Hutchins. Anderson and the Aspen Institute played a crucial role in the launching of a worldwide environmentalist movement, and Earth Day was a big step along the way.

Coincident with the Earth Day effort, *The Progressive,* a 70-year-old publication of the U.S. branch of the Fabian socialist movement of H.G. Wells, Bertrand Russell, and Julian and Aldous Huxley, devoted its entire issue to a special report on "The Crisis of Survival." Among the environmentalist ideologues who contributed to this special issue were Ralph Nader and Paul Ehrlich. Denis Hayes, a Stanford University graduate who would later become the environmentalist-in-residence at the Worldwatch Institute, wrote the keynote article on Earth Day. He stated: "April 22 is a tool—something that can be used to focus the attention of society on where we are heading. It's a chance to start getting a handle on it all; a rejection of the silly idea that bigger is better, and faster is better, world without limits, amen. This has never been true. It presumes a mastery by Man over Nature, and over Nature's laws. Instead of seeking harmony, man has sought to subdue the whole world. The consequences of this are beginning to come home. And time is running out."

In 1970, most Americans would have summarily rejected this pessimistic view. But, by the time the organizers of Earth Day 1970 were planning 20th anniversary celebrations of the event for 1990, the environmentalist hoax had been sold to the population of the United States. In the months before Earth Day 1990, every elementary and secondary school in the nation was provided with a special Earth Day preparation curriculum from the Environmental Protection Agency. EPA spokesmen toured the nation. Television, magazines, and newspapers from the national to local level reported and editorialized on the event. State and town governments promoted it with public funds.

On Earth Day 1990, according to a spokesman for Friends of the Earth (a leading arm of the environmentalist lobby also financed by Robert O. Anderson), "one of the largest demonstrations ever" was held in Washington, D.C., and tens of thousands of people, representing "all types of environmental groups from all over the United States and internationally" were there. Smaller celebrations were held in literally thousands of state capitals, towns, and cities across the United States. A mass movement against science, technology, and economic growth had been consolidated in the United States.

Next Comes Genocide

In 1989, Egyptian President Hosni Mubarak estimated that 500 million people in the Third World had starved to death in the decade of the 1980s; current estimates by the United Nations Children's Emergency Fund (UNICEF) are that 40,000 children under the age of five starve to death every day. Most of these deaths can be attributed directly or indirectly to debt service and "technological apartheid," policies that prevent modern technologies—such as water treatment plants, nuclear energy, refrigeration, mechanized agriculture, pesticides, and fertilizers—from being used in Third World countries. These policies were considered colonialist in past decades; today, they are promoted by environmental groups in industrialized nations, under the guise of saving the Earth from pollution.

Many environmentalists have no idea of the consequences of their belief system for the people of the Third World, but it is clear that those at the top of the environmentalist movement are witting in their advocacy of policies that ultimately kill people. We know this is the case because many of the environmentalist policy-makers say so publicly. It is not simply that the ban on CFCs will kill people and that the top environmentalists know that it will kill people. The fact is that the top ozone depletion propagandists at the World Wildlife Fund, the Club of Rome, the Population Crisis Committee/Draper Fund, and other elite bodies *want* it to kill people. Depopulation is one of the reasons they devised the ozone hoax in the first place. By scaring the general population with stories of imminent catastrophe, these policy-makers intend to justify adoption of stringent measures that will curtail economic growth and population. The ozone hole is just one of several such scare stories.

On July 24, 1980, the U.S. State Department unveiled the *Global 2000 Report to the President.* It had been in preparation by the White House Council on Environmental Quality and the State Department, employing scores of government personnel and hundreds of outside consultants since the early days of the Carter administration—an administration dominated by elite members of David Rockefeller's Trilateral Commission. The report was a long-winded proposal that "population control"—a euphemism for killing people—be made the cornerstone of the policies of all U.S. Presidents from that time forward.

Pervading the report and several companion documents were lurid predictions: crises in water resources, severe energy shortages, shortfalls in strategically vital raw materials—all blamed on "population growth." The report argued that without countervailing action, by the year 2000 there will be 2 to 4 billion people too many. Therefore, the report said, it is required that government implicitly direct all policies domestic and foreign toward the elimination of 2 to 4 billion people by the year 2000.

The rationale for proposing a crime of such great magnitude is the simple—and totally wrong—Malthusian ideology that claims population growth inherently exhausts "natural resources" and that there are, therefore, "limits to growth," as the Club of Rome has insisted.

In the real world of human production of the means of human existence, there is no correlation between "natural resources" and human population potential, for the simple reason that resources are not really "natural." The resources for human existence are defined by human science and technology, and the development of science and technology defines whole new arrays of "resources" for the societies that avail themselves of such progress. For example, oil was there "naturally," but it did not exist as a resource for humankind until the technology—combustion engines, and so on—existed to make it a resource. Before that, it was a black mud that usually meant ruination of farm fields.

This means two things. First, there are no "limits to growth." There are only limits within the confines of a given array of technology. So, unless scientific and technological progress were stopped dead, there could never be an absolute limit to "resources" for human life. There can never be such a thing as absolute "overpopulation" of the human species.

Second, were modern agricultural and industrial capabilities, even as they exist in industrialized nations today, diffused throughout the Third World, we would discover that not only do we have ample resources for year–2000 population levels, but we also have *too few people* to operate advanced agroindustrial facilities at optimum capacity. If we took account of in-sight technological advances, we would discover that *underpopulation* is the main problem we face.

The *Global 2000 Report,* however, assumed no diffusion of modern agroindustrial capabilities to the Third World. Instead, it assumed that the Third World would be denied even available forms of technology. In addition, it assumed no progress beyond existing scientific and technological arsenals. The overpopulation forecast follows neatly from these assumptions: The report assumes that science and technology have been forced to come to a stop, in order to assert that by the year 2000, there will be 2 to 4 billion more people than the world economy can sustain. The report neglects to point out that if science and technology were not to be forced into stagnation, the globe's population would have much brighter prospects.

In other words, the *Global 2000 Report* is simply a statement of a policy intent for genocide, not a scientific forecast at all. It reveals in a unique way the depopulation aims of those also behind the ozone-depletion hoax.

By the time *Global 2000* was issued, whole sections of the U.S. government existed solely to implement its recommendation: depopulation. The role of Richard Elliott Benedick, who negotiated the Montreal Protocol for the United States, must be emphasized again. Benedick has spent most of his government career as head of the State Department Population Office, promoting policies to reduce the size of the world's population.

Lest the skeptical reader think we exaggerate, listen to Thomas Ferguson, a Benedick colleague and head of the Latin American desk at Benedick's Office of Population Affairs. Ferguson made these comments on State Department policy toward the civil war in El Salvador (as reported by *Executive Intelligence Review*, p. 43):

> Once population is out of control, it requires authoritarian government, even fascism, to reduce it. The professionals are not interested in lowering population for humanitarian reasons. ... In El Salvador, there is no place for these people—period. No place.
>
> Look at Vietnam. We studied the thing. That area was also overpopulated and a problem. We thought that the war would lower rates, and we were wrong. To really reduce population quickly, you have to pull all the males into the fighting and you have to kill significant numbers of fertile age females. You know, as long as you have a large number of fertile females, you will have a problem. ...
>
> In El Salvador, you are killing a small number of males and not enough females to do the job on the population. The quickest way to reduce population is through famine, like in Africa, or through disease, like the Black Death. What might happen in El Salvador is that the war might disrupt the distribution of food: The population could weaken itself, you could have disease and starvation. Then you can successfully create a tendency for population rates to decline rapidly. ... But otherwise, people breed like animals.

Ferguson's level of moral depravity is not unique among government policy-makers. Listen to William Paddock, an adviser to the

State Department under both Henry Kissinger and Cyrus Vance. In spring 1981, Paddock told a Georgetown University seminar that 3.5 million of El Salvador's 4 million people should be eliminated, and would be, provided that there was "continuous turmoil and civil strife, which is the only solution to the overpopulation problem."

Paddock continued, "The United States should support the current military dictatorship, because that is what is required. . . . But we should also open up contacts with the opposition, because they will eventually come to power. As we do that, we should work with their opposition, because we will need to bring them to power. That is what our policy is, that is what it must be . . . an endless cycle."

It is no accident that the same "environmentalists" on the consultant list of the *Global 2000 Report* may now be found on the list of sponsors of a ban on CFCs. Nor is it an accident that the ban on CFCs will kill people.

Readers are encouraged to seek out and read the documentation for themselves in official government documents. For example, National Security Study Memorandum 200: *Implications of Worldwide Population Growth for U.S. Security and Overseas Interests,* a recently declassified memo written by National Security Advisers Brent Scowcroft and Henry Kissinger in 1974, states specifically that population growth in the developing sector is a *national security threat to the United States,* and must be curtailed as a matter of America's foreign policy. Under the rubric of this document, the United States has worked internationally to cut the growth and overall size of the darker-skinned peoples of the Third World— an explicitly racist policy.*

*This policy against the Third World and "less advantaged populations" is being implemented on a scale never seen before, but in fact, it is nothing new. Historian Anton Chaitkin documented recently that the policy-makers gathered around George Bush, the family of the President, and the Anglo-American financial establishment behind the Bush administration, are the same group of people who put the racist Adolf Hitler into power and copied his eugenics policies in practice in the United States. They continue to promulgate the policy of Hitlerite "eugenics" or race purification under the new label of population control and in the name of "saving the environment."

Bush's work for population control goes back to the 1960s, when he was the first congressman to introduce national population-control legislation. Bush was also a conspicuous activist for population reduction when he was U.S. ambassador to the United Nations from 1971 to 1972. In 1972, prodded by Bush and others, the U. S. Agency for International Development (AID) began funding the Sterilization League of America to sterilize nonwhites.

References

Peter Brimelow and Leslie Spencer, 1990. "Ralph Nader, Inc.," *Forbes* (Sept. 17), pp. 117–122 (cover story).

Anton Chaitkin and Webster Tarpley, 1992. *George Bush: The Unauthorized Biography.* In Press.

Council on Environmental Quality, 1980. "The Global 2000 Report to the President: Entering the Twenty-First Century," Washington, D.C.

Executive Intelligence Review, 1981. "The Conspiracy Behind the Trilateral Commission," New York (October).

Joseph Farman et al., 1985. "Large losses of total ozone in Antarctica reveal seasonal CLO_x/NO_x interaction," *Nature,* Vol. 315 (Jan. 24), pp. 207–210.

Peter Metzger, 1980. "Government-Funded Activism: Hiding behind the Public Interest." Presented at the 47th Annual Conference of the Southeastern Electric Exchange in Boca Raton, Florida (March 26).

Mario J. Molina and F.S. Rowland, 1974. "Stratospheric sink for chlorfluoromethanes: chlorine atomc-atalysed [sic] destruction of ozone," *Nature,* Vol. 249 (June 28), pp. 810–812.

Kathleen Murphy, 1979. "The 1980s Project: Blueprint for 'Controlled Disintegration,' " *Fusion* (October), pp. 36–47.

National Security Study Memorandum 200, 1974. *Implications of Worldwide Population Growth for U.S. Security and Overseas Interests.* Washington, D.C.

William Paddock, 1981. "The Demographic and National Security Implications of the Salvador Revolution." Washington, D.C.: Georgetown Center for Strategic and International Studies Seminar (Feb. 27).

Sharon Roan, 1989. *Ozone Crisis: The 15-Year Evolution of a Sudden Global Emergency.* New York: John Wiley & Sons, Inc.

Lydia Schulman, 1981. "Global 2000: Will the Zero-Growthers Capture the White House?" *Fusion* (May), pp. 18–19.

In his introduction to the 1973 book *The World Population Crisis: The U.S. Response,* by Phyllis Piotrow, Bush wrote that "one of the major challenges of the 1970s ... will be to curb the world's fertility."

In 1988, U.S. AID made a new contract with the old Sterilization League, committing the U.S. government to spend $80 million over five years. This contract is not listed in the public U.S. AID budgetary literature, yet the group says that 87 percent of its foreign operations are funded by the U.S. government.

The sterilization program is based on deception.

The U.S. AID tells Congress and the public, that since the Reagan and Bush administrations have been opposed to abortions, tax money that would have funded abortions in foreign countries has been diverted to "family planning activities." They fail to explain that in addition to buying 7 billion condoms, the program funds surgical sterilization of growing numbers of the Third World population.

"The State Department's Office of Population Affairs: Depopulating by 'War and Famine,'" 1981. *Fusion* magazine (June), pp. 20–23.i

Nancy Shute, 1983. "The Greening of James Watt," *National Review* (Aug. 5), pp. 924–928 (cover story).

Kathleen Teltsch, 1990. "Rockefeller Foundation Starts Ecology Effort," *The New York Times,* July 24.

11

Gaia: The Environmentalists' Pagan Religion

The Gaia thesis of British ecologist James Lovelock, popular in scientific circles for its concept of the Earth as a living system, has become a quasi-religion for New Age environmentalists, including pagan worship of an Earth goddess and a hatred of the human species that "defiles" her. This chapter will look at the Gaia hypothesis in order to help readers comprehend the belief system of those who are willing to ban a useful man-made substance like CFCs even though there is hard evidence that it will cost human lives and no hard evidence that CFCs are harmful.

Lovelock contends that the Earth is a living being that throughout the ages has influenced and controlled the chemical evolution of its environment to its own benefit. He and his colleague, microbiologist Lynn Margulis, look at the Earth as a biologist might look at a one-celled organism under a microscope, with clouds, rain, and rivers being the planet's circulatory system.

As put forward in four books and many articles by Lovelock, however, Gaia has done more than popularize the notion of the biosphere as a living being. It has become the pseudoscientific robe in which the enviromentalists' naked paganism can be clothed. In the antihuman Gaia-centered view, adopted to varying degrees by most environmentalists, the Earth (or *Gaia,* the name of the ancient Greek Earth goddess) is sacred, while human life is not. Thus, the Earth must be preserved—even at the expense of man.

Lovelock first propounded the Gaia thesis in the 1960s, basing it on work he did for the National Aeronautics and Space Administration (NASA) on how to determine whether life exists on Mars. He found that one guide to the existence of life on Earth is the unique character of Earth's atmosphere, whose composition demonstrates the presence of life. Lovelock also studied the feedback mechanism that prescribes certain boundary conditions—for example, the carbon-dioxide/oxygen balance—which moderates the temperature of the planet.

This scientific aspect of his work is of interest, although it should be noted that much of it is drawn from Louis Pasteur, Pierre Curie, A.V. Gurvitsch, and, especially Soviet Academician V.I. Vernadsky, who coined the term *biosphere* to express the concept that the Earth's surface was composed of an envelope of living matter. However, Lovelock's science is not the basis for the current vogue of the Gaia hypothesis; nor, one suspects, was science Lovelock's actual agenda, even as early as the 1960s. As an active member of the environmentalist community, Lovelock fully subscribes to its axiomatic belief that the human population must be limited in order to protect the continued existence of the biosphere. His latest work, *Healing Gaia,* states this most bluntly in a chapter called "The People Plague." in which he says, "Humans on the Earth behave in some ways like a pathogenic microorganism, or like the cells of a tumor or neoplasm. We have grown in numbers and in disturbance to Gaia, to the point where our presence is perceptibly disabling like a disease." Lovelock calls this "planetary disease" (population, that is) *"Disseminated Primatemia,* the superabundance of humans."

For scientists like Vernadsky or Pasteur, who also considered the question of the biosphere being composed of living matter, the value of extending human life and increasing the world's population was a given. Their concern was how to develop the knowledge and technology to accomplish this mission. In contrast, for Lovelock and today's environmentalists, man is the enemy of the biosphere; man "infests" the planet.

Lovelock's concept of Gaia fills out the primitive environmental-

Matt Moriarty
A Gaia exhibit at the Cathedral of St. John the Divine in New York City, August 1989. The cathedral basement provides a home for many neopagan activities, including the "Temple of Understanding."

ist cosmology by imbuing the Earth with a living personality. The core of this environmental cosmology holds that the human soul is not sacred, nor is the Judeo-Christian God to be worshipped; rather, the Earth itself is personified as a goddess who dominates humankind. The Gaia movement is particularly dangerous because it opens up new and often unsuspecting segments of the population to the neopagan and outright Satanic belief systems that are threatening to overturn the traditional values of Western civilization.

The Birth of Gaia

From fairly unknown beginnings, Lovelock and his Gaia hypothesis began their road to fame at a 1975 conference titled "The Atmosphere: Endangered and Endangering," where leading ecologists gathered to discuss the problems of "overpopulation" and the atmosphere. The conference was organized by New Age anthropologist Margaret Mead, under the sponsorship of the Fogarty International Center and the National Institute of Environmental Health Sciences. Not unexpectedly, participating were the major science personalities who today are promoting hysteria around the global warming issue, including Stephen Schneider, William

*A small selection
of Gaia books.*

Kellogg, and George Woodwell. It was an appropriate setting for Lovelock to launch his Gaia thesis.

The official proceedings of that 1975 conference, published by the U.S. Government Printing Office in 1977, spell out the purpose behind today's calls to shut down modern industries. "The unparalleled increase in human population and its demands for food, energy, and resources is clearly the most important destabilizing influence in the biosphere," Margaret Mead stated bluntly during one conference discussion. The same sentiment was expressed in the first session of the conference, as summarized in the conference proceedings (Kellogg and Mead 1977, p. 73):

> The session was concluded with the thought that we as a species are trying to maintain ourselves at the expense of other species; there seems to be a conflict between preserving nature and feeding the rapidly increasing population. Is our major objective really to feed the population, or do we realize we cannot continue to feed the world *at any price?* Where do we strike a balance between preserving nature and feeding the world?

Mead proposed a global environmental dictatorship to carry out a program of preserving nature, proposing a "Law of the Atmosphere" to regulate relations among states. Her specific statements at this 1975 conference make it clear that the promoters of the ozone hole hoax don't have truth as their main agenda:

> We are facing a period when society must make decisions on a planetary scale. . . . Unless the peoples of the world can begin to understand the immense and long-term consequences of what appear to be small immediate choices—to drill a well, open a road, build a large airplane, make a nuclear test, install a liquid fast breeder reactor, release chemicals which diffuse throughout the atmosphere, or discharge waste in concentrated amounts into the sea—the whole planet may become endangered. . . .
> What we need from scientists are estimates, presented with sufficient conservatism and plausibility . . . that will allow us to start building a system of artificial but effective warnings,

warnings, which will parallel the instincts of animals who flee before the hurricane. . . .

Only by making clear how physically interdependent are the people of all nations can we relate measures taken by one nation to measures taken by another in a way that will draw on the necessary capacities for sacrifice . . . of which human beings—as a group—have proven capable. . . . It is therefore the statement of major possibilities of danger which may overtake humankind . . . on which it is important to concentrate attention [p. xix].

Mead emphatically denounced the solution of development and advanced science and technology, decrying "prophets of paradisiacal impossibilities, guaranteed utopias of technological bliss, or benign interventions on behalf of mankind that are none the less irrational just because they are couched as 'rational.'" Such prophets, she said, "express a kind of faith in the built-in human instinct for survival, or a faith in some magical technological panacea."

Key is Mead's solution, to bring people around to her point of view. This will happen, she said, "only if natural scientists can develop ways of making their statements on the present state of danger credible to each other." Only then, she said, "can we hope to make them credible (and understandable) to social scientists, politicians, and the citizenry."

In a world view like that of Mead, the task for scientists is not to tell the truth, but to make apocalyptic statements that are "credible." The Gaia myth, adopted in whole or even in part, gives the general population an outlook on life that helps them find these scientists' warnings "credible."

Gaia and Man

The Gaia hypothesis is not a matter of science; it is rather an antiscientific, synthetic religion created to unify the environmentalist, women's rights, and peace movements into a New Age movement.

Lovelock makes no effort to hide the mystical character of the Gaia hypothesis. "The concept of Mother Earth, or, as the Greeks called her long ago, Gaia, has been widely held throughout history and has been the basis of a belief which still coexists with the great religions," he writes in his 1979 book, *A New Look at Life on Earth*. "As a result of the accumulation of evidence about the natural environment and the growth of the science of ecology, there have recently been speculations that the biosphere may have been more than just the complete range of all living things within their natural habitat of soil, sea, and air. ... Yet this feeling, however strong, does not prove that Mother Earth lives. Like a religious belief, it is scientifically untestable..." (p. ix).

What role is accorded to human beings by the Gaia hypothesis? Lovelock says in his 1988 book, *The Ages of Gaia*: "The Gaia hypothesis ... focuses special attention on what most people consider to be the lowest part [of life], that represented by the microorganisms. The human species is of course a key development for Gaia, but we have appeared so late in her life that it hardly seems appropriate to start our quest by discussing our own relationship with her" (p. 124).

Lovelock's placement of man in the biosphere is reminiscent of another notorious ecologist with little regard for human life, Adolf Hitler. Hitler wrote in *Mein Kampf* in 1933: "This planet moved for millions of years through the ether without human beings, and it can travel along once again if human beings forget that they owe their higher existence, not to the ideas of some crazy ideologists, but to the knowledge and ruthless application of unshakable natural law."*

In a 1989 interview with the *Orion Nature Quarterly,* Lovelock asserts that there must be a higher force behind the functioning of the universe, and he introduces the Earth goddess at the command center. "There must be something regulating" the oxygen levels on Earth over the course of millions of years, Lovelock told the interviewer. That something, he says, is Gaia, the Earth goddess. He then speculates (West 1989):

> One thing one notices ... in Ireland is the shrines to the Virgin Mary. These shrines are wonderfully tended. They're brightly painted and there are flowers always freshly gathered

---

* Oxford University scholar Anna Bramwell extensively documents that Hitler was an environmentalist in her 1985 and 1990 books (listed in the references). Hitler used technology only as a means of conquest, not as a way to improve the condition of mankind. Today many leaders of Germany's Green Party are former members or admirers of Hitler's S.S.

around them. In contrast, the churches are rather dull ugly places. Maybe these people are not thinking just of the Christian Virgin Mary, but of an older virgin, Gaia, the old Earth Mother.

After all it fits awfully well when you think of it. Gaia is to all intents and purposes immortal. She has lived three and a half thousand million years, which is longer than quite a few stars have lived, and looks like going on for another stellar magnitude age. She is the source of life everlasting. She is certainly a virgin; there is no need to reproduce if you are immortal. She certainly is the mother of all of us in a sense, even Jesus. The whole thing fits as far as Christians go exceedingly well. . . .

And what of mankind's role? In *The Ages of Gaia,* Lovelock (1988) writes:

Gaia, as I see her, is no doting mother tolerant of misdemeanors, nor is she some fragile and delicate damsel in danger from brutal mankind. She is stern and tough, always keeping the world warm and comfortable for those who obey the rules, but ruthless in her destruction of those who transgress. Her unconscious goal is a planet fit for life. If humans stand in the way of this, we shall be eliminated with as little pity as would be shown by the micro-brain of an intercontinental ballistic nuclear missile in full flight to its target [p. 212].

Lovelock, who calls himself "a positive agnostic," then adds:

Gaia is not purposefully antihuman, but so long as we continue to change the global environment against her preferences, we encourage our replacement with a more environmentally seemly species. It all depends on you and me. If we see the world as a living organism of which we are a part— not the owner, nor the tenant; not even a passenger—we could have a long time ahead of us and our species might survive for its "allotted span." It is up to us to act personally in a way that is constructive. The present frenzy of agriculture and forestry is a global ecocide... [p. 236].

Here, explicitly, is the cult religion of Gaia, replacing traditional Judeo-Christian values. Gaia says it is necessary to save Mother Earth even at the expense of individual human life. Thus, stopping the alleged ozone or global warming catastrophe becomes more important than preserving hundreds of millions of human lives. Contrast Lovelock's Gaia and her "preferences" to one of the basic tenets of Judaism and Christianity, as spelled out in the Book of Genesis (Chapter 1, Verses 27–28):

So God created man in his *own* image.... And God blessed them, and God said unto them, Be fruitful, and multiply, and replenish the earth, and subdue it: and have dominion over the fish of the sea, and over the fowl of the air, and over every living thing that moveth upon the earth.

Gaia and Vernadsky

As one might expect, Lovelock took only certain parts of Vernadsky's scientific work, leaving out Vernadsky's insistence that the evolution of living matter proceeds in a definite direction, and that man is the most advanced product of creation. Here is what Vernadsky (1945) writes in "The Biosphere and the Noosphere" (p. 8):

For the first time in the history of mankind the interests of the masses on the one hand, and the free thought of individuals on the other, determine the course of life of mankind and provide standards for men's ideas of justice. Mankind taken as a whole is becoming a mighty geological force. There arises the problem of the *reconstruction of the biosphere in the interests of freely thinking humanity as a single totality*. This new state of the biosphere, which we approach without our noticing it, is the *noosphere*.

The noosphere is a new geological phenomenon on our planet. In it for the first time man becomes a *large-scale geological force*. He can and must rebuild the province of his life by his work and thought, rebuild it radically in comparison with the past. Wider and wider creative possibilities open

273

before him. It may be that the generation of our grandchildren
will approach their blossoming. . . .

The important fact is that our democratic ideals are in tune
with the elemental geological processes, with the laws of
nature, and with the noosphere. Therefore we may face the
future with confidence. It is in our hands. We will not let it
go.

In other words, in order to carry out the actual laws of nature,
man has to increase the free energy of the biosphere. If man does
not do that, then we will suffer a biological holocaust caused by the
biosphere's entropic collapse. How do we increase free energy?
Simply by ensuring the continued development and application
of more advanced technologies that increase the energy density
per unit area. The course of human progress demonstrates how
this has occurred. Primitive societies, for example, burned wood
for fuel. As technology advanced and wood became scarce, new
energy sources were developed—fossil fuels, nuclear, and next
fusion, each orders of magnitude more energy dense. Thus the
Gaia enthusiasts who promote "appropriate technologies" and
"sustainable development"—which translates into wood burning
and deforestation in the Third World—are actually causing the
very ecocide they decry by insisting on less energy-dense fuels.

The Popularization of Gaia

Many readers might dismiss Gaia as nonsense. However, the
Gaia hypothesis has taken such hold in the scientific community
that the American Geophysical Union, one of the world's leading
scientific associations, sponsored a four-day conference on Gaia
in San Diego in 1988. The March 7–11 conference featured an
impressive list of scientists who presented the pros and cons of
the Gaia hypothesis from diverse perspectives.

Another indication of Gaia's fame is a 15-page glossy paean by
the *Encyclopedia Britannica* titled "Gaia: A Goddess of the Earth?"
in its 1988 *Yearbook of Science and the Future*. Its author, Stephen
Schneider, is best known as a publicist for the greenhouse doomsday scenario. Schneider, who also spoke at the 1975 Fogarty International Center conference, has taken over editorial control of
many scientific journals relating to climate, from his position at
the National Center for Atmospheric Research in Colorado. "To

some the idea of planetary-scale homeostasis, the principal intellectual thrust behind the Gaia hypothesis, is more like religion than science. As religion Gaia can be deep, beautiful, and fascinating," Schneider wrote in the *Yearbook*.

Gaia has even been the subject of several television specials, such as Public Broadcasting's NOVA series, which popularized it as "Goddess of the Earth." And Anchor Press/Doubleday has published a glitzy 272-page atlas, titled *Gaia: An Atlas of Planetary Management,* which has become a bible for ecologists and is being used in schools and universities. The foreword announces that the atlas "show[s] how we are plundering our planet in the most profligate and dangerous way. . . . [A]ll the ills that beset us . . . can be traced back inexorably to three root causes: overpopulation, political stupidity, and wasteful misuse of the planet's treasures, both finite resources, and renewable living wealth."

Who Is Behind Lovelock?

Lovelock likes to portray himself as a quiet scientist who works in the isolation of his Devonshire, England, country home, in touch with Gaia. However, he is part of an organized fellowship, the Lindisfarne Association, that actively organizes and promotes its Gaia ideology.

In the basement of the Episcopal Cathedral of St. John the Divine on West 110th Street in New York City is the Gaia Institute, as well as the Temple of Understanding and other cult and Satanic operations. The Gaia Institute, according to Priscilla Peterson, the director of the Temple of Understanding, aims to create "mother goddess" cults throughout the West. As Peterson put it, the Gaia Institute is involved heavily "in the ecological battle of preserving the Earth . . . a movement to create a new religion. . . . There is certainly no question that great goddesses or God in a feminine form has been worshipped by a lot of societies not just in Antiquity, but more recently."

The philosophy behind Gaia is put forward in *Gaia, a Way of Knowing,* published in 1987 by the Lindisfarne Press (Thompson 1987). The book is a compilation chiefly of papers presented by James Lovelock and other Lindisfarne Fellows at their 1981 conference at the San Francisco Zen Center. Gaia's purpose, the book says, is "to create a new ecology of consciousness, the basis for a new political and economic order which, because it arises out of the study of life, is life-enhancing and life-embracing."

In addition to Lovelock and Lynn Margulis, the other contributors are all familiar names of the New Age movement today. The conference and the book were both orchestrated by William Irwin Thompson; who cofounded Lindisfarne along with Gregory Bateson (the British operative of MK-Ultra whose special project was to introduce psychotropic drugs—LSD, in particular—to the flower children of the 1960s), and his former wife, anthropologist Margaret Mead. Bateson's contribution to the book is a transcript of a tape Bateson made in 1980, shortly before his death, giving his last instructions to the Lindisfarne Association.

The Gaia "way of knowing" has an international network with one of its coordinating centers in Scotland at the Findhorn Foundation, described by some as the "Vatican City" of the New Age movement. Certain of Findhorn's trustees, like Ed Posey of London, are also codirectors of Britain's Gaia Foundation. The Gaia Foundation of Britain is controlled and patronized by individuals closely linked to the British Royal Family. One of its chief advisers, Sir Laurens van der Post, is the mentor of Prince Charles on matters of mysticism and philosophy. One of its official patrons, Lady Geraldine Ogilvy, is the niece-by-marriage of Royal Family member Angus Ogilvy, husband of Princess Alexandra. Another patron of the Gaia Foundation is Sir John Harvey-Jones, former chairman of Imperial Chemical Industries and a former British intelligence operative in the Soviet Union (according to his official biography). The ICI connection is interesting, because ICI, like Edgar Bronfman's Du Pont Company, is a top funder of the movement to ban CFCs (see Chapter 9).

The Gaia Foundation houses an organization called Forest People's Support Group and the Education of the Awakening Earth. It sponsored top Brazilian ecologist José Lutzenberger for a Gaia organizing tour of Britain in autumn 1988. Lutzenberger at that time headed the Gaia Foundation in Brazil. While he was the environment minister of Brazil, Lutzenberger was one of the recipients of the Right Livelihood Award, presented by the Gaia Foundation and its collaborators to individuals for their "vision and work contributing to making life more whole, healing our planet, and uplifting humanity." Recipients are chosen by an international jury panel, all of whom also serve as directors of the Right Livelihood Foundation. These jurors have included: Rodrigo Carazo, former president of Costa Rica, founder and president of the U.N. Peace University; Monika Griefahn, Greenpeace; and Robert Muller, former assistant secretary-general of the United Nations and an important international collaborator of the Lucis Trust, an elite neo-Satanic organization that operates the Temple of Understanding at the United Nations and funds many environmentalist groups. (The Lucis Trust was originally called the Lucifer Trust, but its name was changed after protests at the United Nations from several Catholic nations. Its purpose has not changed, however, and it is currently organizing for the 1992 Earth Summit in Brazil.)

Today, Lutzenberger is a leading organizer of the 1992 Earth Summit, which will gather together representatives of the world's governments—including numerous heads of state—in Rio de Janeiro, under the auspices of the United Nations Conference on Environment and Development (UNCED).

Gaia Is Not Science

On March 8, 1990, in an article bearing the title "Hands Up for the Gaia Hypothesis," the noted British science magazine *Nature* gave Lovelock three full pages in which to present the Gaia hypothesis, which the magazine billed as a major "scientific discovery" (pp. 100–102). Doubtless Lovelock chose his title with the image in mind of humankind raising its hands in adoration of Gaia, the Earth Mother. His "Hands up!" however, signals an assault on science, in particular in its role as a driver for the progress of Western civilization.

To the objection that the biosphere is only a small part of the Earth, and that there is no need to resurrect Mother Earth from Greek mythology, Lovelock counters with a "profound" new definition of the concept of life. We characterize a tree as "living" without hesitation even though the interior wood is dead and only the outermost layer under the bark is "living" biologically, he says. So why not say the Earth is living, and that the biosphere produced the "wood" of the Earth's core?

By *Nature*'s standards, it may be that the Gaia hypothesis is science. However, as measured against the standards that have marked the course of the history of modern science since the Renaissance, Lovelock's hypothesis certainly is not science. Geologist James W. Kirchner makes this point emphatically in a May 1989 article in *Reviews of Geophysics,* "The Gaia Hypothesis: Can It Be

277

Tested?" Kirchner says that Lovelock's Gaia hypothesis can be neither proved nor disproved as a scientific hypothesis. It is, in reality, not a scientific hypothesis.

This criticism must have stung Lovelock to the core, for in his "Hands Up!" (1990) article, he compares Kirchner to an agent of the Inquisition. Kirchner defends himself in a letter to *Nature* published June 7, 1990, in which he reemphasizes that a scientific hypothesis must be provable, and that the Gaia hypothesis is not. Much of the scientific data that Lovelock uses to bolster the Gaia hypothesis is contradictory, Kirchner writes:

> For example, Lovelock proposed that algae act as a global thermostat, producing Dimethylsulfide (DMS), a precursor of cloud condensation nuclei, to try to cool the Earth when it is warm. Unfortunately, ice core data published the next year suggested exactly the opposite. DMS levels are apparently highest during glacial periods; if the algae are important at all, they act to make the Earth even colder when it is cold. Undeterred, Lovelock promptly inverted the theory to match the data. He now argues that Gaia actually prefers the glacial deep freeze and that the interglacial periods represent merely "a fevered condition of the planet" [p. 470].

Despite such sound criticism, there is increasing enthusiasm in the scientific community to create a scientifically reputable existence for Gaia. Gaia-devotee Stephen Schneider, for example, has provided special services in this connection, for which Lovelock properly honors him in the 1990 *Nature* article and elsewhere. Schneider gave a boost to the Gaia hypothesis as a "scientific breakthrough" with a paper at the 1988 San Diego conference of the American Geophysical Union. Prior to that, according to insiders, there were only five scientists who paid any attention to the Gaia hypothesis; subsequently, the number of scientific papers on Gaia climbed into the hundreds.

Readers who think that we are overestimating the importance of the modern-day Gaia proponents should consider the following: The multinational IBM corporation is contributing grant money for the introduction of school children on five continents to the "theory and practice of Gaia" through the Song of Gaia. And what is this Song of Gaia? It is a hymn celebrating the goddess Mother Earth, which originated at a festival given by the Commonwealth

Institute of London annually at Christmas for children from the 49 countries of the Commonwealth. Here are the words of the hymn:

Gaia is the one who gives us birth. She's the air, she's the sea, she's Mother Earth. She's the creatures that crawl and swim and fly. She's the growing grass, she's you and I.

References

Anna Bramwell, 1985. *Blood and Soil: Richard Walther Darré and Hitler's 'Green Party.'* Bourne End, England: The Kensal Press.

——, 1990. *Ecology in the 20th Century.* Oxford: Oxford University Press.

William W. Kellogg and Margaret Mead, eds., 1977. *The Atmosphere: Endangered and Endangering. A Conference Sponsored by the Fogarty International Center for Advanced Study in the Health Sciences and the National Institute of Environmental Health Sciences, National Institutes of Health, Oct. 26–29, 1975.* DHEW Publication No. (NIH) 77–1065. Washington, D.C.: U.S. Government Printing Office.

James W. Kirchner, 1989. "The Gaia Hypothesis: Can It Be Tested?" *Reviews of Geophysics,* Vol. 27 (May), pp. 223–235.

——, 1990. "Gaia Metaphor Unfalsifiable," *Nature,* Vol. 345 (June 7), p. 470.

James E. Lovelock, 1979. *A New Look at Life on Earth.* Oxford: Oxford University Press.

——, 1988. *The Ages of Gaia: A Biography of Our Living Earth.* New York: W.W. Norton & Co.

——, 1990. "Hands up for the Gaia hypothesis," *Nature,* Vol. 344 (March 8) pp. 100–102.

——, 1991. *Healing Gaia: Practical Medicine for the Planet.* New York: Harmony Books.

Norman Myers, ed., 1984. *Gaia—An Atlas of Planet Management.* Garden City, N.Y.: Anchor Press, Doubleday & Co.

Stephen H. Schneider, 1987. "Gaia—A Goddess of the Earth?" In *1988 Yearbook of Science and the Future.* Chicago: Encyclopedia Britannica, p. 29.

William Irwin Thompson, ed., 1987. *Gaia—A Way of Knowing: Political Implications of the New Biology.* Great Barrington, Mass.: Lindisfarne Press.

V. I. Vernadsky, 1926. *Biosfera* (The Biosphere). Leningrad: ONTI. (A French edition, *La Biosphere,* was published in Paris by Alkan in 1929.)

——, 1944. "Problems of Biogeochemistry, II—The Fundamental Matter-Energy Difference Between the Living and the Inert Natural Bodies of the Biosphere," *Transactions of the Connecticut Academy of Arts and Sciences,* Vol. 35 (June) pp. 483–517.

————, 1945. "The Biosphere and the Noosphere," *American Scientist,* Vol. 33 (January), p. 6.

Ross Evan West, 1989. "Gaia—She's Alive: A Conversation with James Lovelock," *Orion Nature Quarterly,* Vol. 8, No. 1, p. 58 (published by the Myrin Institute, Inc., New York, N.Y.).

12

Great Projects to Transform the Globe

A tragic result of the ozone depletion hoax is that the hysteria created by the environmentalists' manufactured threats has masked very real global problems that threaten mankind and the environment today. In this final chapter, we will discuss these actual biological and ecological disasters. The underlying "disease" that has caused these problems is not man, as the environmentalists allege, but poverty. Therefore, we present a series of great projects that can be launched to reverse the present descent into an ever-deeper economic depression. We do this very specifically to give readers who are angry about environmental hoaxes a positive alternative to fight for—not just for the United States, but for the entire world. It is this broad view of developing the world that must be counterposed to the smallness and meanness of the environmentalists' remedies.

One of the most serious ecological problems today is the threat of *global deforestation*. In Africa, deforestation has led to an expansion of the Sahara and Kalahari deserts and to the degradation of large tracts of land. In other parts of the world, deforestation is causing profound shifts in climate. (One of the first articles on this issue appeared in *21st Century Science and Technology* magazine in January–February 1989, authored by Rogelio Maduro.)

Environmentalists claim that logging is the leading cause of deforestation. Actually, the cause is the lack of modern energy sources and the poverty that accompanies it. More than 60 percent of global deforestation is the result of the use of wood as a fuel source. A study by the United Nations has documented that 83 percent of logs cut down around the world are used as firewood. Another 20 to 25 percent of deforestation is the result of slash-and-burn primitive agriculture. Commercial logging accounts for approximately 18 percent of deforestation, and most logging companies replant trees to cover the area cut down. Therefore, a halt to commercial logging will not significantly stop global deforestation.

The solution is advanced energy production, including nuclear energy, and modern agricultural production methods—exactly the opposite of what the environmentalists propose (they advocate the use of firewood as a fuel source).

Poverty, together with the collapse of morality, is also the root cause of the second major ecological holocaust facing the world today: *epidemic disease*. A worldwide collapse in the public health infrastructure and living conditions over the past two decades has led to a resurgence of previously conquered global pandemics and the appearance of new ones. AIDS, cholera, tuberculosis, and malaria are all spreading out of control. In most countries of Africa, the infection rate for the human immunodeficiency virus (HIV), which causes AIDS, now ranges between 30 percent and 50 percent of the population. There is no cure for AIDS, a 100 percent lethal disease, which means that tens of millions of Africans will die in the next few years of complications from AIDS. The situation in the United States is not much better. Heterosexual transmission has now become the main mode of transmission of AIDS, and Black, Hispanic, and Native American teenagers are the majority of those now being infected. The federal government's Centers for Disease Control sentinel hospitals in the Bronx and other major urban areas report rates of HIV infections as high as 20 to 30 percent. In other words, the U.S. inner cities already have levels of AIDS infection as high as those in Africa.

The world health crisis is burgeoning: at the same time that 70 million Americans have little or no medical insurance, more than 100 hospitals close down every year because of a lack of funds. The drug epidemic is out of control, destroying and enslaving the minds of the young, and with that the future of those nations that abet and tolerate it. The collapse of public services, to pay for usurious interest rates, continues at a increasingly rapid pace, as even the U.S. government faces bankruptcy.

A third threat is the *rapid collapse of the economies of Eastern Europe,* a collapse that is will result in a series of bloody wars, especially in the former Soviet Union, unless present World Bank and International Monetary Fund policies are reversed.

These are the real ecological disasters facing mankind today, not the imagined threat of ozone depletion and global warming.

A Solution: American System Economics

The development of public health policies, medical science, and economic prosperity created a health revolution, the result of which is that many of the world's people can expect to live to the age of 70 or older. By looking at the example of those policies that shaped this change, we can achieve the same kind of recovery.

In the United States, these policies were articulated by Alexander Hamilton, the first secretary of the treasury, and they became known as the American System, a term coined by the great American statesman Henry Clay. Crucial to the economic program of Hamilton, and later Abraham Lincoln, was the investment of finance capital in the development of infrastructure and manufacturing. Within a short period of time after the implementation of Hamilton's policies, as embodied in his 1792 *Report to the Congress on the Subject of Manufactures,* the young republic was transformed from a bankrupt debtor nation to a prosperous, independent country with no foreign debt.

Unlike today's environmentalists, the American System economists believed that "people are wealth." In other words, they defined the wealth of a nation as the creative potential of its citizens. To maximize this creative potential, it was necessary to ensure conditions for the working population to have adequate wages to provide their families with a nutritious diet, suitable housing and health care, and a top-rate education.

The remedy today, to reverse the environmental and biological holocausts now facing the world, is the same as that proposed by Hamilton 200 years ago. Nations must adopt the American System of political economy of America's Founding Fathers.* This chapter

---

* For those who wish to learn more about the American System, we recommend *The Harmony of Interests,* written in 1851 by Henry C. Carey, who later became

discusses the kind of great development projects that can achieve this goal.

The World Needs More People

The history of mankind has been shaped by the large-scale development of infrastructure, from the Grand Canal with its vast irrigation systems in ancient China, to the rise of the railroad and canal systems in the United States and Europe in the middle of the 19th century, to take just two examples. These developments completely changed the "ecosystems" of vast areas not for the worse but in such a way that a greater density of human population could be maintained.

This is the law of life on the planet Earth, which was true long before the emergence of man: Life constantly transforms its environment, thereby creating the conditions for an ever-growing density and intensity of the living process per unit land area.

Today, the world stands on the threshold of a new era of great infrastructure projects, projects that are absolutely necessary in order to maintain a population of more than 6 billion persons, growing to 12 billion around the middle of the 21st century. These great projects will serve as the basis for transforming and uplifting the economy of the globe, making it possible for a growing population to live at standards as high as or higher than those of the United States during the decade that the Apollo "great project" to land on the Moon was pumping wealth into the U.S. economy.

The notion that the world is overpopulated by human beings is a retread of the discredited thesis of the 18th century's Parson Thomas Malthus, who was hired by the colonialists at the British empire's East India Company to counter the optimistic economic growth policies being implemented by the new United States republic. Malthus lied that mankind's ability to produce the wealth needed to sustain the species will always be outstripped by the biological growth of mankind's numbers.

The truth is that the world needs *more people*. Global economic development is hindered by the lack of population in most of the Earth's land area. More than half of the area of South America

---

President Abraham Lincoln's economic adviser. Carey's books and those of other American System economists, including Peshine Smith, Mathew Carey, and Friedrich List, are available from Augustus Kelley Publishers in New York City.

has less than one inhabitant per square kilometer. The average population density of Africa is one-eighteenth that of West Germany! Most of Canada, the great West of the United States, most of the Soviet Union with its vast land mass spread over 11 time zones, the Amazon River basin of South America, the vast interior of Australia, and the deserts of the Near East are nearly empty of human population. Although we are reaching the capability to colonize Mars, the colonization of the Earth has really only begun. If all nations of the world had the population density of central Europe, the Earth's population would be more than 35 billion.

Presented here is a partial list of great infrastructure projects that have already been proposed and worked out; they await only the political will to set them into motion. These great projects will require enormous numbers of machine tools, trucks, bulldozers, cranes, and other equipment that could be supplied by the United States. In fact, the demand would be so great that instead of massive unemployment rolls, the United States would face labor shortages. All together, these projects will employ directly tens of millions of workers, but for each worker so employed, the lives of more than one hundred people will be transformed and improved.

For those truly concerned with improving the environment on Earth, these great projects are the place to start.

THE EUROPEAN 'PRODUCTIVE TRIANGLE'

The rapid collapse of the economies of Eastern Europe and the former Soviet Union is creating the conditions for a series of wars that cannot be contained. To reverse this disaster, as well as the worldwide economic collapse, requires that the Great Projects for world economic reconstruction begin in West-Central Europe, which has the capability now to create the capital goods, upgraded manpower, and other resources needed for the global effort. American economist and statesman Lyndon H. LaRouche, Jr., has put forward a program that seeks to use the advanced technology and skilled labor resources of West Central Europe as the generation point for a general economic recovery throughout the West, and rapid industrialization of the Third World. This productive triangle, as it is known, is a triangular region approximately the size of Japan with sides connecting Paris, Berlin, and Vienna. At its center is the construction of an integrated, computerized high-speed rail system for freight and passenger transport throughout

THE EUROPEAN PRODUCTIVE TRIANGLE

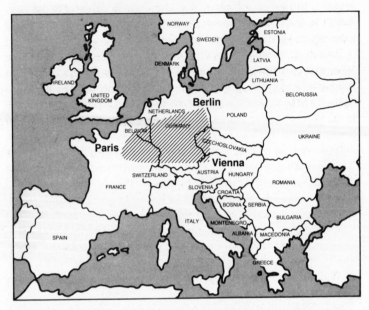

A triangular region approximately the size of Japan, created by a transportation grid connecting Paris, Berlin, and Vienna, must become the generation point for Western economic recovery and rapid industrialization of the Third World. This highly developed and densely populated region can "jump-start" the U.S. economy with exports of machine tools and other industrial inputs no longer produced in the United States—in much the same way that America jump-started Europe's economy after World War II. Nuclear power and maglev transportation technologies will be featured.

the region. By focusing investment into this and a few other crucial areas of industrial infrastructure, the Paris-Berlin-Vienna triangle can be transformed into the greatest concentration of productive economic power the world has ever seen.

Radiating outward from this central triangle will be great infrastructure corridors linking it to the entirety of Europe. One arm reaches out from Berlin to Warsaw, and down to the Polish industrial center of Krakow-Katowice; another reaches up through Hamburg to Scandinavia; another via Paris to the rest of France and Spain; another via Basel into Italy; another via Vienna to Budapest, and so on. Like a giant spiral galaxy, this system of a dense center with outreaching arms of economic development, encompasses a total market of more than 430 million persons in Western Europe, eastern Germany, Czechoslovakia, Hungary, and Poland. It is the key to the economic reunification of Europe—to restoring the continental European economy as a coherent cultural and infrastructural unity. It is also the key to bringing the developing sector into the 21st century and reindustrializing the United States, reestablishing the tremendous productive capacity that is now dormant or destroyed.

For the efficient transport of bulk goods (raw materials, steel, chemicals, grain, and so on) two main infrastructural improvements are required:

(1) Modernization of existing medium-speed rail systems, including a total overhaul of the systems in the eastern region of Germany and northern Czechoslovakia. Emphasis is to be placed upon renewing East-West rail connections that were disrupted by the division of Europe, including emphatically the corridors running from the northern Ruhr region via Braunschweig to Berlin, via Kassel-Göttingen-Nordheim into Leipzig, and further to the south from Frankfurt via Erfurt. In addition, attention should be given to the Munich-Prague and Vienna-Prague corridors.

(2) Completion and improvement of the central European river and canal system to make it fully usable by 1,350-ton Europa-class ships.˙ Once the long-planned Elbe-Danube connection is constructed, the center of this system will form a closed ring: It will run from the Ruhr region via the Midland canal to Magdeburg and Berlin; via Magdeburg on the Elbe southward via the planned Elbe-Oder-Danube canal, with connections to Prague on the Mⁿˑ dau, to Bratislava; then up the Danube via Vienna connecting the Rhine-Main-Danube canal near Regensburg, continuing oⁿ

SPIRAL ARMS OF THE TRIANGLE
Great infrastructure corridors of modern communications, high-speed railways, canals, and industrial "nuplexes," will draw a total market area of 430 million people and thousands of small, high-technology businesses, into an economic development era that will unify Eastern and Western Europe.

Main down to Frankfurt, and from there down the Rhine back to the Ruhr region.

Cheap and plentiful energy—above all, electricity—is the second key to an "economic miracle" in the Paris-Berlin-Vienna triangle. The model to be followed is France's nuclear energy program as originally planned under Charles de Gaulle. Not only does nuclear energy provide France with electricity at half the cost of alternative sources, but its use can eliminate the terrible pollution caused by heavy use of brown coal in the eastern region of Germany and other East bloc countries. The most urgent task is to rebuild the energy systems of the eastern side of the triangle—that is, the eastern region of Germany and western Czechoslovakia. For this purpose, beyond immediate measures to modernize coal power plants that still have useful life, nuclear energy is the only viable solution. Construction of an additional 6 billion watts (gigawatts or GWe) of nuclear power in the eastern region of Germany, and 4 GWe in western Czechoslovakia should be begun as soon as possible.

For a variety of reasons, the modular high-temperature gas-cooled reactor (HTGR) is the most attractive design for the eastern and central areas of the triangle. This reactor type is intrinsically the safest in existence—it shuts itself down by physical mechanisms and is incapable of producing a "meltdown." HTGR modules of about 150 MW, as designed by the General Atomics Company in San Diego, can be produced in assembly-line fashion and installed in 27 months or less. They can provide high-temperature industrial process heat as well as electricity and lower-temperature heat for district heating. Their inherent safety characteristics permit them to be installed directly inside urban industrial centers.

Finally, modernization of telecommunications systems in the eastern region of Germany and western Czechoslovakia must proceed at the highest technical standards required by the central triangle as the world's most advanced industrial region of the year 2000 and beyond. The logical step is to go directly to the digitized radiotelephone system already projected for Western Europe ("D-system"). Within five years or so, people will be able to communicate with each other anywhere in central Europe, without interruption: at home, at work, in automobiles and trains.

New economic opportunities are also to be found in such badly needed projects as the construction of a bridge over the Gibraltar Strait linking Spain to Morocco, a bridge from the Italian mainland to Sicily, the reconstruction of transport infrastructure throughout the former Soviet Union and Eastern Europe, and revival of the old "Baghdad-Bonn" railroad from central Europe through the former Yugoslavia and Bulgaria, to Istanbul and through to Baghdad, with connections to Iran and also via the Eastern shore of the Mediterranean to North Africa.

GREAT PROJECTS FOR THE UNITED STATES
Reconstruction of America's Basic Infrastructure

The United States must begin paying a gigantic accrued "infrastructure bill" of trillions of dollars. We must rebuild our crumbling inner cities, highways, bridges, water supplies, and power systems—the essential infrastructure that has been suffering from neglect and under-investment for the past 25 years, and is now on the verge of collapse. An economics team from *Executive Intelligence Review* magazine estimated that during the past 18 years, the nation has so depleted its basic economic infrastructure, by lack of repairs, that today's repair bill, for restoring quality of infrastructure, per capita and per acre, to 1970 levels of functional quality would be approximately $4 trillion.

We outline here two great projects for the United States.

NAWAPA: The North American Water and Power Alliance

One of the great projects proposed 30 years ago in detail would provide the continent, and particularly the water-starved U.S. West, with both power and water—the North American Water and Power Alliance or NAWAPA. The best place to look for solving the water needs of North America is the most abundant region of surplus fresh water flowing on the face of the Earth—the northwestern region of America, where upwards of a quarter of all the rain and snow that hits land on this planet cascades down rivers out of Alaska and the Canadian Rockies, untouched by anyone, northward into the Arctic Ocean, or westward into the Northern Pacific.

By reversing the flow of 15 percent of this water, turning it southward and eastward, the continent could benefit from as much as 180 million acre-feet of new fresh water per year. The Pasadena-based Ralph M. Parsons engineering company prepared the original plans for NAWAPA in the early 1960s. The idea involved dam-

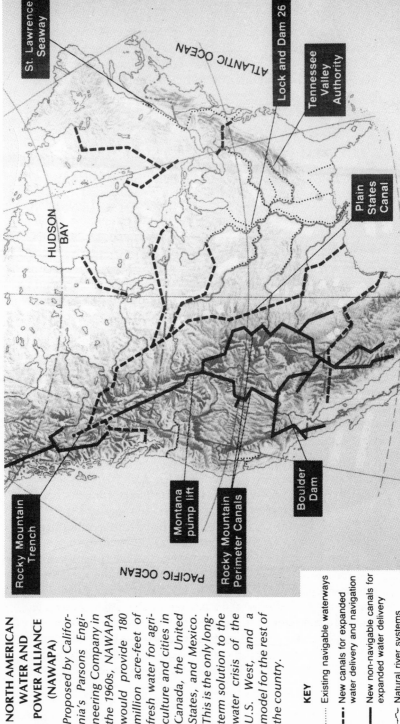

NORTH AMERICAN WATER AND POWER ALLIANCE (NAWAPA)

Proposed by California's Parsons Engineering Company in the 1960s, NAWAPA would provide 180 million acre-feet of fresh water for agriculture and cities in Canada, the United States, and Mexico. This is the only long-term solution to the water crisis of the U.S. West, and a model for the rest of the country.

St. Lawrence Seaway

Lock and Dam 26

Tennessee Valley Authority

Plain States Canal

Rocky Mountain Trench

Montana pump lift

Rocky Mountain Perimeter Canals

Boulder Dam

ATLANTIC OCEAN

PACIFIC OCEAN

HUDSON BAY

GULF OF MEXICO

GULF OF CALIFORNIA

KEY

- .... Existing navigable waterways
- ▪–▪–▪ New canals for expanded water delivery and navigation
- ▬▬▬ New non-navigable canals for expanded water delivery
- ∼∼∼ Natural river systems

ming three rivers in Alaska and Canada's Yukon Territory and directing the water southward via a chain of reservoirs, dams, and channels, making use of the 500-mile Rocky Mountain Trench, located at an elevation of 3,000 feet in the Canadian Rockies. Only in Montana would lift pumps be needed, to move the water higher so that the gradient would be enough for gravity flow to continue, even as far as Mexico.

Such a project could produce, in addition, a massive surplus of hydroelectric power, the ability to maintain constant water levels on the Great Lakes and Mississippi River, and new navigable waterways linking the Great Lakes to the Pacific Ocean across the Canadian prairie, and the Canadian prairie provinces to the Mississippi.

Abundant water to recharge the depleted Ogallala Aquifer, upon which 22 million acres of irrigated cropland now depend in Texas, New Mexico, Colorado, Oklahoma, Kansas, and Nebraska, would be available, as well as enough to relieve the overtaxed Colorado River of its burden to provide the water needs for southern California, Arizona, and Mexico. Also, 22 million acre-feet of water would be available for the northern agricultural states of Mexico, including augmentation of the flow of the Rio Grande River.

New water resources would be ample to allow bountiful farming, meeting residential and industrial needs, and creating barge transport along some rivers that are not navigable. Benefits would flow to 23 states.

NAWAPA is designed to link with regional water systems development, including desalination plants in California.

The 'New American Railroad': A Maglev Train Network

High-speed modern magnetically levitated (maglev) trains, which can travel more than 300 miles per hour, should replace highway transport and especially air routes for the most heavily traveled corridors on both coasts of the nation, and then extend to the Mississippi River, and beyond to the metropolitan centers of Texas and the Great Plains states. The route most in need of maglev technology is the heavily congested Northeast corridor, where both highways and airports are operating at beyond peak traffic loads, and accident deaths are increasing dramatically as a result. The map shows a conceptual two-phase plan for connecting hub airports to high-speed maglev ground transport arteries. If the United States undertakes funding and development of such a

NEW AMERICAN RAILROAD

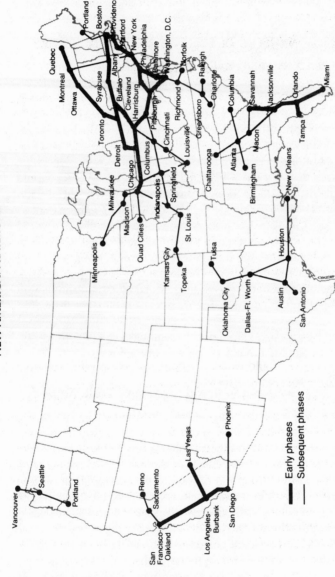

Early phases
Subsequent phases

A network of high-speed magnetically levitated trains to relieve congestion of major highway systems and airports, especially in the Northeast and the U.S. industrial heartland, and bring the U.S.A. up to par with Japanese and German transportation technologies.

project immediately, the nation has the capability to leapfrog into second and third generations of maglev technologies, catching up and even surpassing Germany and Japan, which are now ahead in the development of this 21st-century mode of transport.

GREAT PROJECTS OF THE SOUTHERN HEMISPHERE
Water Projects

All of Latin America from Mexico to Cape Horn would benefit greatly from an integrated approach to the development of the region's water resources. Consider the alternative: During the decade of the 1980s, the dismantling and cancellation of water resource projects throughout South America—at the behest of the World Bank and International Monetary Fund—has created the conditions for today's epidemics of deadly diseases, led by the cholera epidemic spilling over the borders of Peru.

A fully integrated approach is needed, such as that pioneered 50 years ago by the United States Tennessee Valley Authority (TVA), which transformed much of the rural American Southeast from dirt farming to modern industrial agribusiness. The TVA not only dammed the Tennessee River for hydroelectric power to feed local industry, but also built a large number of smaller dams to control flooding and established a series of environmental agricultural stations that today stand as leaders in research in fertilizers and high-yield agriculture.

Five projects would create the basis for regionwide economic development of South America:

• the Plhino and Plhigon water projects in Mexico, with two grand North-South canals, 1,100 and 1,400 kilometers (about 685 and 870 miles) in length.

• the complex of projects for utilizing the entire La Plata River Basin, including improvements of the Paraná River-Paraguay River and connection of that water system of more than 3,000 kilometers in length, made fully navigable, to the Amazon system, in part by the planned Guapore-Paraguay canal. This would make possible massive irrigation projects in the fertile de la Plata region.

• two canal projects linking rivers in Brazil to the Atlantic coast.

• the trans-Andean water pumping projects in Peru.

• the opening up of the Llanos area of Colombia and Peru by appropriate water works.

GREAT WATER PROJECTS FOR IBERO-AMERICA

Shown here are the sites of major canals and tunnels to be constructed during a 30-year water management master plan, which will supply the continent with adequate fresh water supplies for industry, agriculture, and domestic use. Two plans for Mexico, the Northeast Water Plan and the North Gulf Water Plan, are not shown.

1. Second Panama Canal
2. Atrato-Truandó Canal
3. Orinco-Rio Negro Canal
4. Madeira-Guaporé Canal
5. Lake Mamoré-Guaporé
6. Guaporé-Paraguay Canal
7. Chaco Canal
8. Bermejo Canal
9. Lake Iberá
10. Tieté Canal
11. Ibicui-Yacui Canal
12. Andean tunnels

RIO DE LA PLATA WATER PROJECTS

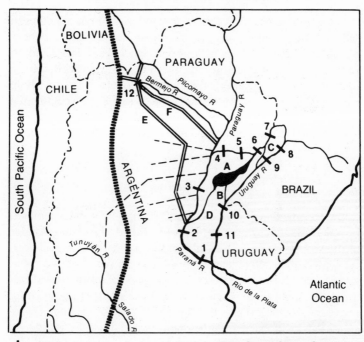

I dams	**⊪⊪⊪** Grand Chaco Canal
– – proposed aqueducts and canals	**═══** Canals of Bermejo River

A Lake Iberá
B Mirinay River
C Aguapey River
D Corrientes River
E Santiago del Estero Canal
F Lateral Canal

1 Guazú
2 Paraná Medio Chapetón

3 Parará Medio Pati
4 Itatí
5 Yacycretá Compensador
6 Yacyretá
7 Corpus
8 Roncador
9 Garabí
10 San Pedro
11 Salto Grande
12 Zanja del Tigre

A major Ibero-American water development site is the complex of projects for utilizing the entire La Plata River Basin, including improvements of the Paraña River-Paraguay River and connection of that water system of more than 3,000 kilometers (1,860 miles) in length, made fully navigable, to the Amazon system, in part by the planned Guapore-Paraguay canal. This would make possible massive irrigation projects in the fertile de la Plata region.

Modern Transportation

In addition to water projects, South America requires a modern rail grid. This would include the completion and improvement of a transcontinental railway from Rio de Janeiro and Saõ Paulo, to Santa Cruz and La Paz, through the Andes to the Pacific Ocean, opening up the rich mineral resources of Bolivia and of the whole region. This and other priority rail projects to achieve a continental rail system in South America, plus a high-performance rail line from Colombia through Central America to Mexico, must be built.

A Second Panama Canal

The Panama Canal, started by the French in 1879 and completed by American engineers in 1914, is often called "the eighth wonder of the world." Today the canal handles some 14,000 vessels carrying more than 160 million tons of cargo per year. But its capacities are severely overtaxed, even in this period of relative slowdown in international trade. Furthermore, today's largest ocean-going vessels are too big to pass through the canal's lock system. Under conditions of global economic recovery and the industrial development of Asian and Ibero-American nations, the canal would be unable to handle any of the new sea-going traffic generated. A second route from Atlantic to Pacific is badly needed.

The optimal plan for a second Panama Canal would involve the construction of a two-way, sea-level canal crossing the isthmus from the Chorrera district of Panama on the Pacific to a point near the mouth of the Lagarto River on the Caribbean. The canal could be finished within 12 to 14 years at a cost of $15 billion to $18 billion, and would allow the simultaneous passage in opposite directions of two ships of a little more than 300,000 tons each. Its construction will require the biggest excavation in history, employing 10,000 workers in direct construction work and another 5,000 in subsidiary work.

Two industrial complexes should also be built, one at each end of the new canal, to transform domestic and foreign raw material into finished and semifinished goods for internal consumption and export. For example, Cerro Colorado copper could be turned into wire and electrical motors.

Not including the entrance passages on the Caribbean and Pa-

RAILWAY GRID FOR IBERO-AMERICA

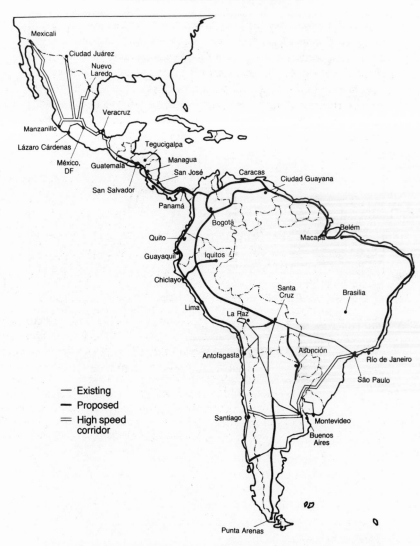

This master plan includes the completion and improvement of a transcontinental railway from Rio de Janeiro and Sao Paolo, to Santa Cruz and La Paz, through the Andes to the Pacific Ocean, opening up the rich mineral resources of Bolivia and of the whole region. The goal: to achieve a continental rail system in South America, plus a high-performance rail line from Colombia through Central America to Mexico.

THE SECOND PANAMA CANAL

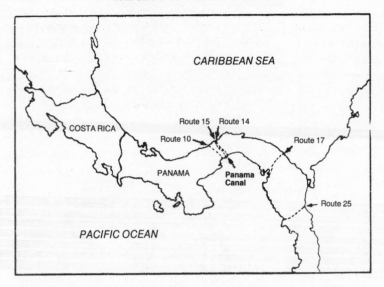

The Panama Canal is overcrowded and obsolete; it is too small for today's largest sea-going vessels. A new sea-level canal is needed, and could be built at any one of the proposed sites shown on the map, depending on construction technology used. The new canal could generate sufficient income to pay for itself in no more than 30 years.

cific, the total length would be about 82 km (about 51 miles), and the crossing through the land part would take only about six hours.

The sea-level canal will generate sufficient income to pay off the entire debt in no more than 30 years. This money will not come from Panama's national treasury but from the canal itself.

GREAT PROJECTS FOR ASIA

Developing the Indian and Pacific Ocean Basins

By the year 2000, close to two-thirds of the world's population will live in the countries on the Pacific and Indian Ocean rim. If these countries are to survive and grow in the next decade, they will have to take deliberate steps to restart and hasten the process of industrialization throughout the region, which has been slowed by the global depression. The best way to accomplish this is through investment in great projects of infrastructure.

There are three major infrastructural projects required to permit efficient commerce into and within the Indian Ocean and Pacific Ocean basins. One has been discussed above, the sea-level canal through the Panamanian isthmus, and to this we add discussion of a second major canal project, the construction of a canal through the isthmus of Thailand. The third great transportation project affecting the Indian and Pacific Ocean basins, not considered here directly, is the improvement of the Suez Canal. The additional three proposed projects are principally inland water-management development projects: (1) the north-south Grand Canal modernization project within China; (2) the Mekong River development project for Southeast Asia; and (3) the water development project for the vast fresh-water potentialities of the Indian subcontinent, principally India, Pakistan, and Bangladesh. Also discussed below is a more elaborate plan for the infrastructural development of China, the world's most populous country, put forward in the 1920s by Dr. Sun Yat-sen, the founding father of China.

The Kra Canal

As far back as 1793, the younger brother of King Rama I of Thailand (then Siam) proposed, for military and commercial reasons, to dig a canal across the peninsula of Thailand south of the Isthmus of Kra, connecting the Lake of Songkhla and the South

THE KRA CANAL

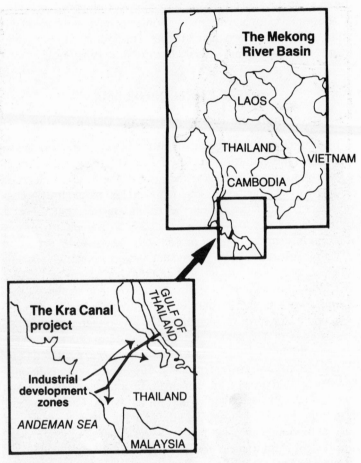

This project, first proposed nearly 200 years ago in 1793, would connect the South China Sea with the Indian Ocean. It would relieve growing congestion at the Straits of Malacca past Singapore, and create vast industrial development potential based on construction of deep sea ports at one or both of the canal outlets.

THE MEKONG CASCADE

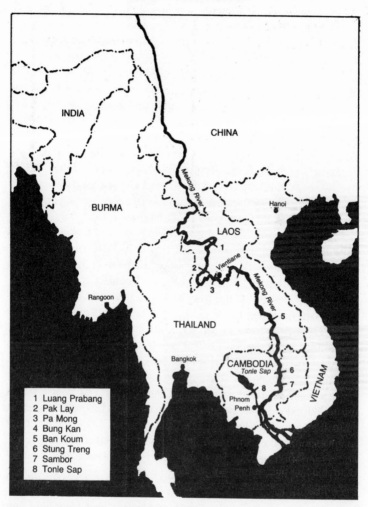

Control of the Mekong River and development of the Mekong Delta could create a new breadbasket in Southeast Asia. The Mekong Cascade, an integrated system of dams and reservoirs, has been studied since 1956. The plan envisions the construction of eight dams and reservoirs and five major power projects, at a cost of about 20 billion 1990 dollars—approximately 4 percent of the annual take from the world drug trade.

303

China Sea with the Indian Ocean. A modern feasibility study for the Kra Canal was commissioned in the early 1970s by Mr. K.Y. Chow of the Thai Oil Refining Company, and prepared by a U.S. engineering firm in collaboration with the Lawrence Livermore National Laboratory in 1973.

The actual savings in distance traveled effected by a Kra Canal—about 1,448 km (900 miles)—would not by themselves justify the large expenditures on excavation and operating costs. There are two other principal factors that define the overall importance of the project: the growing inadequacy of the Straits of Malacca past Singapore; and the industrial development potential based on construction of deep sea ports at one or both of the canal outlets. An "Asiaport" conjoined with the Kra Canal could become a major trade center for Southeast Asia, capable of eclipsing Singapore. In addition, it has the potential of serving as a focal point for the development of massive shipbuilding, heavy construction, and capital-goods manufacturing industries. It would be of major strategic significance to all of Southeast Asia and would function as a unifying element in the national life of Thailand.

The 1973 design for the Kra Canal specifies a total canal length of 102 km (64 miles), with sea approaches of 50 km (31 miles) in the West and 70 km (44 miles) in the East, respectively. This is the shortest possible route for a sea-level canal that bypasses mountainous terrain and does not pass into the territorial waters of Burma. The canal is designed to accommodate tankers of at least 500,000 deadweight tons, accommodating a draft of 110 feet (fully loaded supertankers). It is to have two channels, to permit East-West and West-East traffic simultaneously. Construction time by conventional excavation methods is estimated at 10 to 12 years. Excavation by use of peaceful nuclear explosives (PNEs) would cut both construction time and cost by 40 percent.

Development of the Mekong Delta

The Mekong River is one of the world's largest rivers and is Southeast Asia's greatest single resource. Four riparian countries form the watershed of the Mekong River: Cambodia, Laos, Vietnam, and Thailand. The lower Mekong Basin covers more than 609,000 sq km (380,625 sq miles) comprising almost the whole of Laos and Cambodia, one-third of Thailand, and two-fifths of southern Vietnam. Virtually the entirety of the water resources of this huge river now goes unutilized, flowing through the Mekong Delta into the South China Sea. The development of the Mekong Delta is the essential infrastructural project to unlock the industrial and agricultural potential of these nations.

Today the lower Mekong drainage basin has a population of approximately 50 million, representing more than one-third of the total population of the four riparian countries of Laos, Cambodia, Thailand, and Vietnam. At average per capita annual incomes of $100 to $350, the basin area's people, including those in rural northeastern Thailand, are among the world's poorest.

The Mekong is the world's eighth largest river in annual flow. Along its course, the Mekong drains a total catchment area of 795,000 sq km (307,173 sq miles)—well over twice the land area of Japan—including some 185,000 sq km (71,436 sq miles) along its upper reaches in China and Burma. The average annual discharge is more than 475,000 million cubic meters (387.4 trillion acre-feet) of water, with large seasonal variations.

The tremendous power potential of this river can be precisely specified. Based on total average annual discharge into the sea, runoff from various parts of the basin, and differences in elevation between the center of each unit surface and the mean sea level, the theoretical hydroelectric potential has been calculated to be 58,000 megawatts of installed capacity and 505,000 gigawatt-hours of annual energy generation. Of this, 37,650 and 194,000 GWh per year have been found to be technically feasible by latest studies. By comparison, Thailand's total present electricity consumption is only about 40,000 GWh, or 20 percent of that generation potential.

Plans for the development of the delta have been prepared by the United Nations Economic Commission for Asia and the Far East (1947), the U.S. Bureau of Reclamation (1956), which recommended the construction of five mainstream multipurpose dams; and the Mekong Committee, with representatives from Laos, Cambodia, Thailand and Vietnam (1970). The current program for river development is today known as the Mekong Cascade, an integrated system of dams and reservoirs that would regulate the lower 2,000 km of mainstream flow of the numerous systems studied since 1956. The plan, put forward by the Interim Mekong Committee, envisions the construction of eight dams and five major power projects. The total project costs would be in the range of $20 billion in 1990 dollars—about 4 percent of the annual take from the illegal world drug trafficking.

Development of the Mekong Delta could create a new bread-basket in Southeast Asia. A United Nations study of the late 1960s estimated that rice production could be increased from 12 million to 37 million tons per year. Similar increases were projected for production of livestock, fish, and fibers. These are now considered to be very conservative estimates, reflecting a low projected level of mechanization and fertilization.

India's National Water Regulation Project

A study conducted in 1979 by the Fusion Energy Foundation—still applicable in its essentials today—shows India's capacity to become a modern industrialized nation if the latest, most advanced technologies are used. The key to Indian economic development is the huge task of harnessing the subcontinent's gigantic water resources, to break the centuries-old cycle of droughts and floods that has stood in the way of modern agriculture.

According to the foundation's study, the development of India's water resources would cost $180 to $200 billion (in 1979 dollars) over a 30-year period. It would quadruple India's electricity-generating capacity from 8,000 to more than 40,000 megawatts. The study also showed that India's grain production, which was at 120 million tons per year in 1979, could be increased to somewhere between 1 and 2 billion tons per year. With the necessary inputs of fertilizer and mechanization, India could become a breadbasket for the world, with more than 120 million irrigated acres.

The 30-year plan involves three stages of development, entailing the construction of a network of dams, reservoirs, canal systems, and nuclear-industrial centers known as nuplexes. The major components of the plan include a Brahmaputra-Ganges Canal; completion of the Rajasthan and Ganges-Rajasthan canals; the Ganges-Cauvery and Link canals from south India, linking with the Brahmaputra-Ganges canal in Madhya Pradesh; the regulation of all rivers and large-scale irrigation; and the construction of "nuplex centers" combining nuclear energy and agroindustries, including fertilizer production.

The North-South Grand Canal of China

In the early 1980s, work began in China on the modernization of an ancient system of canals joining China's greatest river, the

INDIAN SUBCONTINENT WATER CONTROL PROGRAM

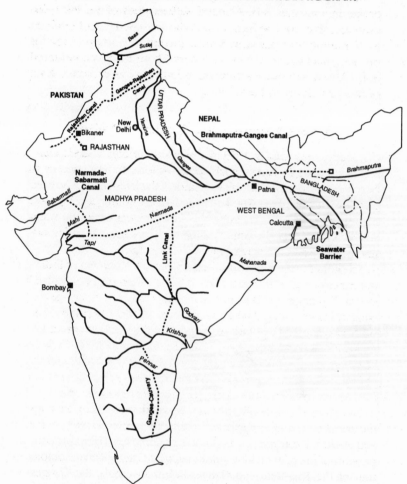

During the 1970s, the Fusion Energy Foundation developed a continentwide 30-year program to control and harness India's vast water resources, with dams, reservoirs, canals, nuplexes, and hydroelectric plants. It would break the centuries-old cycle of droughts and floods that has slowed modernization of agriculture, and would quadruple hydroelectric production of electricity for industry.

NORTH-SOUTH GRAND CANAL IN CHINA

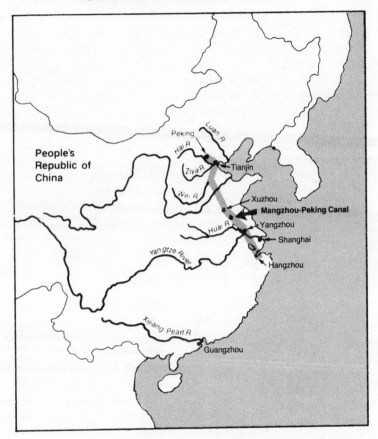

This north-south water diversion project centers on the modernization of the famous Grand Canal, an ancient waterway over which grain taxes were once shipped to the northern imperial capitals from the grain-producing regions of the South. This canal could now play a major role in facilitating modern transportation within China, which has had historic problems with north-south transit, because most of the country's rivers flow east to west.

Yangzte, with the Yellow River and other rivers and lakes of northern China. This north-south water diversion project centers on the modernization of the famous Grand Canal, an ancient waterway over which grain taxes were once shipped to the northern imperial capitals from the grain-producing regions of the South. This canal could now play a major role in facilitating modern transportation within China, which has had historic problems with north-south transit, because of the fact that most of the country's rivers flow from east to west.

The primary objectives of the project are to increase transport capacity, particularly for the movement of coal and other essential raw materials; water diversion from the Yangzte River to the northern cities suffering critical water shortages; and to provide, in conjunction with a network of secondary canals and river systems, a capacity for flood control and irrigation in the parched agricultural regions of the north China plain.

Dr. Sun Yat-sen's Infrastructure Program

Sun Yat-sen (1866–1925) was an ardent nationalist and an avowed follower of Abraham Lincoln. He is considered by the Chinese as the founding father of modern China. Like Lincoln, Dr. Sun realized that in order to build a nation, the country's infrastructure—its land, water, education, and transportation systems—had to be built up. He believed that only under such conditions could China become a great economic and cultural power. In his memoirs, published in 1927, Sun Yat-sen summarized the program he proposed for China's national reconstruction and for which he sought international investment.

"If the program is gradually carried out," Dr. Sun wrote, "China will become, not a mere 'dumping ground' for foreign goods, but a real 'economic ocean,' capable of absorbing all the surplus capital of the world as rapidly as the industrial countries can produce."

Dr. Sun's program featured the development of modern systems of transportation, including 100,000 miles of railways, 1,000,000 miles of roads, the improvement of existing canals; construction of two new canal systems; the organization of China's river system; clearing and deepening the bed of the Yangtze, from Hankow to the sea, in order to permit ocean-going vessels to reach Hankow; clearing and deepening the bed of the river Hwangho, to prevent flooding; clearing the Hsikiang, the Hwaiho, and other rivers;

SUN YAT-SEN'S RAILWAY SYSTEM FOR CHINA

The infrastructure program developed by Dr. Sun Yat-sen to bring China into the 20th century included waterways, communications, and energy technologies, but its focus was the development of a national transportation grid: at least 100,000 miles of railways, complemented by a million miles of roads.

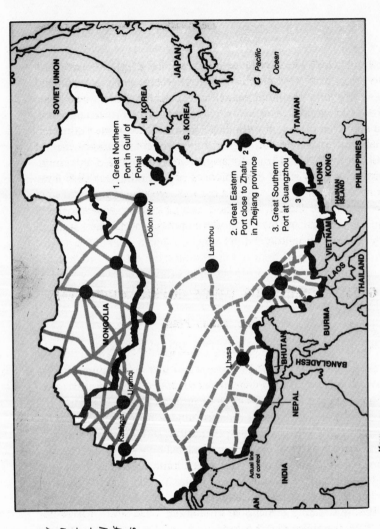

Key:

Part of the Northwestern Railway System suggested by Dr. Sun Yat-sen to open up virgin lands of Xinjiang and Mongolia

Part of the Southwestern Railway System suggested by Dr. Sun Yat-sen to integrate Xizang (Tibet) and southern Xinjiang with the southern part of China

1. Great Northern Port in Gulf of Pohai
2. Great Eastern Port close to Zhafu in Zhejiang province
3. Great Southern Port at Guangzhou

SOVIET UNION

JAPAN

N. KOREA

S. KOREA

Pacific Ocean

TAIWAN

HONG KONG

HAINAN ISLAND

PHILIPPINES

VIETNAM

LAOS

THAILAND

BURMA

BANGLADESH

BHUTAN

NEPAL

INDIA

MONGOLIA

Dolon Nor

Lanzhou

Lhasa

Urumqi

Kashgar

Actual line of control

construction of long-distance telegraph and telephone lines, and also organization of wireless telegraph stations; organization and development of commercial harbors; organization of three large-scale ocean ports, capable of equaling New York City in the future, in the north, center, and south of China; construction of commercial and fishing harbors along the entire coast; construction of commercial docks along all navigable rivers; the building of modern cities, with social conveniences of all kinds, near all railway centers, principal stations, and harbors; erection of iron and steel works on the largest scale, and also of cement works to meet building requirements; and development of China's mineral wealth and agriculture.

GREAT PROJECTS FOR AFRICA AND THE MIDDLE EAST

The 'Oasis Plan'

The Oasis Plan for regionwide development of the Middle East, a project commissioned by economist Lyndon LaRouche, centers on solving the desperate fresh-water supply problem in arid regions of north Africa and the Middle East. As LaRouche has discussed, the Oasis Plan is also the key to peace in the Mideast, providing areas of collaboration for mutually beneficial projects to green the deserts. The Oasis Plan includes:

(1) Completion of the Jonglei Canal improving water use of the White Nile.

(2) Diversion of water from the Ubangi River in Zaire via canals, pipes, and pump stations to refill Lake Chad and provide for massive irrigation of the Sahel.

(3) Construction of a sea-level canal from the Mediterranean to the Qattara Depression in Egypt, combined with power generation, large-scale desalination and regional development.

(4) Engineering and construction of navigable canals from the Mediterranean to the Dead Sea through Israel and from the Dead Sea to the Red Sea through Jordan, with nuclear plants and desalination facilities along the way.

(5) Building large-scale, nuclear-powered desalination centers generating "artificial rivers" of fresh water for irrigation and consumption, for old and new population centers and industries to create "green belts" in North Africa and the Middle East.

OVERVIEW OF OASIS PLAN FOR THE MIDDLE EAST

This Great Project features the construction of two major pipelines: one through Israeli territory from the Dead Sea to the Mediterranean, and one through Jordan from the Dead Sea to the Red Sea. Along the banks of these navigable canals, nuclear-powered desalination complexes would generate "artificial rivers" of fresh water for irrigation and consumption.

1 Dead Sea Canal
From the Mediterranean Sea to the Dead Sea, for desalination, transportation, and irrigation

2 Red Sea Canal
A second canal through Jordan from the Dead Sea to the Red Sea

3 Qattara Depression
Canal from the Mediterranean to the Qattara Depression: create a lake, build hydroelectric and desalination plants along canal

4 New Lakes and Rivers
Engineered by man, throughout the Middle East region

5 Advanced nuclear power technologies
For desalination, construction, industry, and urban power needs

6 Jonglei Canal plan
Create an efficient water channel through the upper White Nile swamp; create thousands of acres of prime farmland; add water to the Nile River

7 Ground Water Development
Utilize large underground aquifers in the Sahara, Egyptian deserts, and northeastern Saudi Arabia

8 Lake Chad-Congo Basin Development
Create a "Great Lakes" of central Africa, providing water for agriculture, transportation, power, sanitation, industry, and beauty for the continent

THE DEAD SEA CANAL

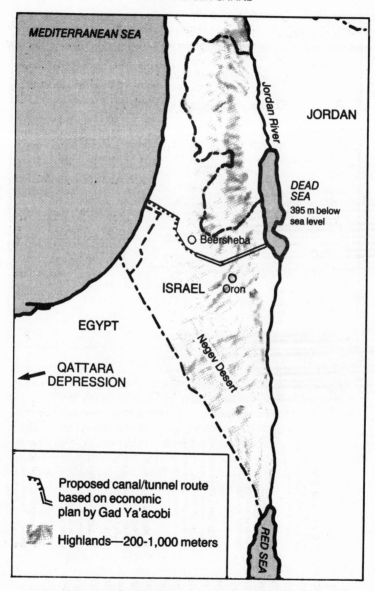

Also proposed for the Middle East is engineering and construction of a canal from the Mediterranean to the Dead Sea, with nuclear plants and desalination facilities along the way. A similar canal could be built from the Dead Sea to the Red Sea, passing through Jordan.

THE QATTARA DEPRESSION

About 50 miles south of the Mediterranean Sea, in the western desert of Egypt, there is a dry, massive depression below sea level, known as the Qattara Depression. Plans for cutting canals from the sea and flooding this area have existed for decades. Water would be delivered through the canals by pump, and hydroelectric turbines installed along the canal are estimated to be capable of generating about 10 gigawatts of hydroelectric power. Long-term benefits of the project include a marked change for the better in local climate, due to increased evaporation of water, and the medium-term greening of the desert.

THE JONGLEI CANAL

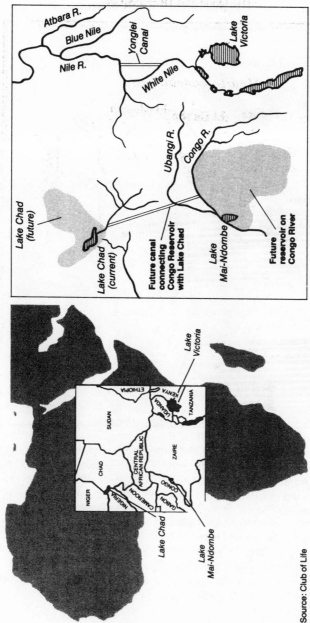

Source: Club of Life

In southeastern Sudan, where the upper White Nile rises before joining the Blue Nile and flowing as the Nile River into Egypt, there are extensive marshy areas known as the Sudd swamp. Construction of a canal at Jonglei southward to Malakal would drain the swamps and save precious water now lost to evaporation. Hundreds of thousands of acres of prime farmland would be created. This project was begun but halted by funding problems.

THE AFRICAN CENTRAL LAKE PROJECT

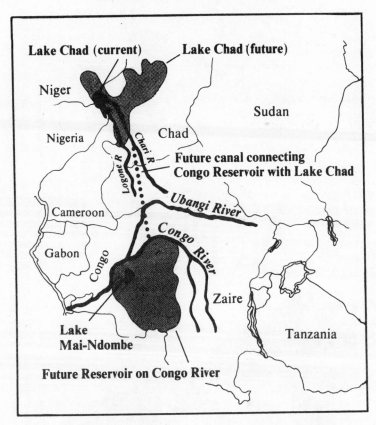

In 1982, the Global Infrastructure Fund plan released by Japan's Mitsubishi Institute proposed the taming of the Congo River in northwest Africa by damming the river north of Brazzaville, to create a vast new lake—about 130 square miles in area—in the Congo and Chad regions of central Africa. A new canal would transport some of the new water resources to enlarge Lake Chad in Central Africa. This immense project would create water resources for the irrigation of 800,000 square miles of land, where 110 million people live. The new Congo Lake would be connected with secondary rivers by a system of canals, which would become an axis for regional interior transportation. The hydroelectric potential of the region would be doubled.

Okavango-Zambezi Water Project in Southern Africa

The Great Projects needed for southern Africa center on power and water. South Africa has plenty of the former—primarily coal-generated—and a scarcity of the latter. South Africa's neighbors need energy, but have plenty of undeveloped water resources. An aggressive regional development of water and hydropower resources is the key not only to economic growth, but also to lasting peace in the region.

The map shows a water resources development plan for southern Africa developed by South African water engineer Desmond Midgley in 1984, excerpts of which were published by *Executive Intelligence Review* magazine in 1990. The plan involves pumping water from the Zambezi River on the border between Zambia and Namibia, at Katimo Mulilo, into a major canal that would carry it south through Botswana, and, via a series of pump stations, canals, and aqueducts into the water-short agricultural and ranching areas of eastern Botswana and northern South Africa's Transvaal. Discharge from the main canal would augment the Boteti River in Bostwana and the Limpopo River, which forms the northern border of South Africa with Botswana. Branch lines would supply Gaborone, the capital of Botswana, and other growth points.

With 1,340 km (833 mi) of canals, and 80 km (50 mi) of tunnels, pipelines, and siphons, the Zambezi-Transvaal link would constitute one of the world's largest water transfer projects. It would supply needed fresh water at one-seventh the capital costs and one-sixth the operating costs projected for desalination plants along South Africa's ocean shores.

A Trans-African Rail System

Also crucial to the rapid development of the dying continent of Africa is the construction of a network of modern railways. A two-phase construction program, developed by the Fusion Energy Foundation in Germany in 1978, is shown on the map.

The heart of a solution to the urgent needs of the continent as a whole is a rail system cutting across the Sahel, preferably from Dakar in the west to Djibouti in the east. The east-west rail line (actually projected as early as the 1870s) intersects the existing railway systems of Nigeria and Egypt-Sudan. The rail network must then be extended by north-south intersecting trunk lines: north-

OKAVANGO AND ZAMBEZI RIVER WATER DIVERSION PROJECT

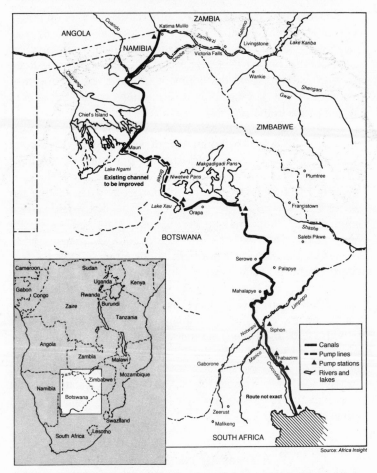

Source: Africa Insight

Southern Africa can be rapidly developed, by combining the technological know-how of South Africa with the enormous natural resources of the southern subcontinent. One bottleneck to industrial and agricultural progress that must be tackled, however, is lack of fresh water supplies for energy production, irrigation, and manufacturing. Water resources will be diverted southward from the Zambezi River on the border between Zambia and Namibia, and the Okavango River in Botswana, through a network of pump stations, gravity cannels, tunnels, and aqueducts. This approach is much cheaper than proposed plans for large-scale desalination of ocean water.

AN AFRICAN RAIL NETWORK

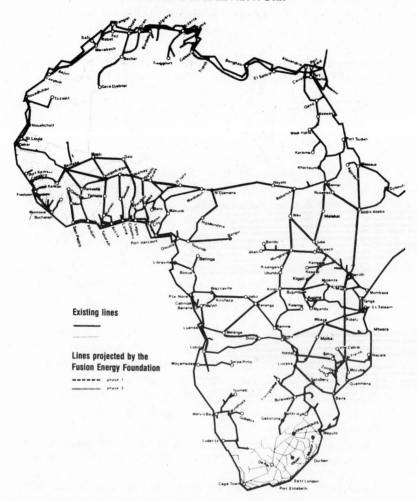

The pattern of modern-day rail lines in Africa demonstrates that the railways were built by former colonial powers, for serving their military needs and for the cheap transportation of raw materials to port cities for transport to the home country. In 1982, the Fusion Energy Foundation proposed a continentwide rail development program, which would begin with construction of an east-west line from Djibouti to Dakar, linked with upgraded and newly constructed north-south lines across the continent.

Source: Fusion Energy Foundation

319

south from Algeria and south into Tanzania. The east-west rail line across the Sahel serves as the indispensable logistical baseline for beginning to reverse the present spread of the Saharan desert into the Sahel.

By joining this with north-south links, this rail system will play a vital part in fostering initially modest but crucial trade among the nations of both Arab and black Africa. The combined effects of railway development and water management are optimal exploitation of combined railways and navigable waterways, creating the beginnings of a functioning internal transport system. This development, enriched by the beginnings of the development of a continent-wide system of energy production and distribution, provides the logistical base to begin the in-depth improvement of agriculture throughout the continent.

It Can Be Done!

Many Americans believe—erroneously—that we cannot afford to finance essential development projects like those discussed here. So let us ask the question: How will these great projects be financed?

In brief, the answer is that the creation of new markets in the Third World and in Eastern Europe is the key to revived prosperity in the industrialized economies. By supporting and even underwriting these projects, the United States and its allies would be creating new markets and new wealth in the developing sector. Within a short time, these great projects would begin to pay for themselves.

With our work on these great projects here on Earth under way, man can then raise his eyes to the Heavens and begin the great project of space exploration and colonization begun in the 1960s, but sadly aborted. Eventually we shall carry out great projects on Mars, to transform it from a dead desert into a beautiful garden, fit for billions of human beings.

What honest nature-lover could object to that?

References

Executive Intelligence Review, 1991. *Can Europe Stop the World Depression?* (Special Report, June 9) Washington, D.C.

Marsha Freeman, 1991. "A TVA on the Jordan," *21st Century Science & Technology* (Spring), pp. 45–46.

Fusion Energy Foundation, 1980. "A 40-year Plan: Making India an Industrial Superpower," *Fusion* (May), pp. 35–48.

Global 2000: Genocide 100 Times Worse Than Hitler, 1981. New York: National Democratic Policy Committee.

Marjorie Mazel Hecht, 1990. "MHTGR: Nuclear Power to Develop the World," *21st Century Science & Technology* (Spring), pp. 28–35.

Lyndon LaRouche, Jr., 1983. "A Fifty-Year Development Policy for the Indian-Pacific Oceans Basin," *Executive Intelligence Review* (Special Report, August), Washington, D.C.

————, 1991. "LaRouche's 'Oasis Plan': The Only Real Basis for Mideast Peace," *21st Century Science & Technology* (Spring), pp. 38–40.

Rogelio A. Maduro, 1989. "The Climatic Consequences of Razing the Rain Forest," *21st Century Science & Technology* (January–February), pp. 26–35.

Marcia Merry, 1991. "Man-made Rivers and Lakes Can Transform the Mideast and Africa," *21st Century Science & Technology* (Spring), pp. 40–44.

D.C. Midgley, 1990. Water and Electric Power to Develop Southern Africa," *Executive Intelligence Review,* Vol. 17, No. 11, March 9, pp. 34–37.

Ralf Schauerhammer, 1991. "An Outline of the Productive Triangle," in *Can Europe Pull Us Out of Economic Collapse?* Washington, D.C.: Democrats for Economic Recovery, LaRouche in 92, pp. 18–26.

Jonathan Tennenbaum, 1990. "Sun Yat-sen's Grand Design for Industrializing China," *21st Century Science & Technology* (Spring), pp. 40–47.

Appendix

This appendix is reprinted from Chapter 6 of *Exploring the Atmosphere* by G.M.B. Dobson (1968) by permission of Oxford University Press.

6. OZONE IN THE ATMOSPHERE
1. Introduction

As we briefly mentioned in Chapter 1, although there is very little ozone in the atmosphere, it is of interest for three quite different reasons. First, since ozone absorbs ultraviolet light very strongly, a large part of the ultraviolet radiation from the Sun (about 5 percent of the total energy received by the Earth from the Sun) is absorbed by the ozone high in the atmosphere, and this energy heats the air at this height and is the cause of the warm region at a height of about 50 km above the ground. Second, the amount of ozone in the atmosphere above any place is found to vary considerably from day to day and, although most of the ozone is situated above the tropopause, yet these variations are found to be closely associated with variations in the weather. Finally, in addition to the day-to-day variations, the ozone has a very unexpected type of annual variation, and its distribution over the world is equally unexpected. From these variations with season and latitude, it has been possible to deduce something about the general worldwide circulation of the air between the troposphere and the stratosphere, and this has recently become of special interest in view of the fall-out of radioactive matter from nuclear explosions.

2. General Characteristics of Ozone

There are three different forms of oxygen in the atmosphere and for the benefit of the reader who has but little knowledge of chemistry we may distinguish between them as follows.

Oxygen atoms In this, the simplest form of oxygen, the atoms exist as separate units. Much of the oxygen in the highest atmosphere exists in this form.

Oxygen molecules In this form each molecule contains two atoms of oxygen which are bound together. This is the normal form in which oxygen exists in the lower atmosphere. If separate oxygen atoms existed in the lower atmosphere, they would very rapidly combine in pairs to form oxygen molecules.

Ozone Each molecule of ozone contains three atoms of oxygen.

A. *Chemical Properties of Ozone*

While ozone is a form of oxygen, its properties are very different from those of normal oxygen. One of its three atoms is easily given up to any substance with which it comes into contact and which is readily oxidizable, and the ozone is thereby destroyed. A large amount of energy is required to form ozone from oxygen and this energy is given up again when ozone returns to normal oxygen; pure ozone is actually explosive and will detonate under suitable conditions. On the other hand, a large amount of energy is required to dissociate a molecule of oxygen into two single atoms of oxygen and this energy is released again when two atoms of oxygen recombine to form a molecule of oxygen.

B. *Physical Properties*

For our present purpose the chemical properties of ozone are of less importance than its physical properties, the most important of which is its ability to absorb radiation in certain parts of the spectrum. Oxygen is almost transparent to the whole of the visible, infrared, and ultraviolet parts of the spectrum, absorbing only the rather short wavelengths, less than 2,400 Å. On the other hand, ozone absorbs radiation in three different parts of the spectrum.

(*a*) It absorbs very strongly in the ultraviolet region between about 3,300 and 2,200 Å, the strongest absorption being at a wavelength of about 2,500 Å where it is extremely intense. The small amount of ozone in the upper atmosphere effectively cuts off the whole of the incoming radiation from the Sun of wavelengths less than 3,000 Å. It is this absorption of solar radiation that is the cause of the formation of the warm region in the atmosphere at a height of about 50 km that was described in Chapter 3 (see Figure 3.6). It is also of interest to note that, if it were not for this absorption, the inhabitants of the Earth would be subject to excessive sunburn. (*b*) Ozone absorbs radiation in the yellow-green region of the spectrum. Though the absorption here is rather weak, yet, because it occurs in a part of the spectrum where the amount of incoming solar radiation is at its maximum intensity, it gives rise to appreciable warming in the upper atmosphere. (*c*) The third important region of absorption by ozone is in the infrared part of the spectrum, chiefly at wavelengths around 9.5 μm (95,000 Å). This absorption band of ozone happens to come in a part of the spectrum where the outgoing radiation from the Earth and the lower atmosphere is at its maximum. It has been suggested that the absorption of this outgoing terrestrial radiation leads to some warming of the stratosphere, but the theoretical treatment of the subject is very difficult and it is still uncertain whether this warming is important or not.

C. *Formation and Destruction of Ozone*

When the oxygen in the upper atmosphere absorbs solar radiation of wavelengths shorter than about 2,400 Å, molecules of oxygen are broken up into atoms of oxygen. When one of these atoms of oxygen encounters another molecule of oxygen it may combine with it to form a molecule of ozone. It is believed that almost all the ozone in the atmosphere is formed in this way. The rate at which ozone is formed in the upper atmosphere will naturally be greatest at the levels where there are plenty of both oxygen atoms and oxygen molecules and the main region of formation is at heights above 30 km. There is, however, another process going on at the same time; when ozone absorbs solar radiation of wavelengths around 2,500 Å, it is itself destroyed, forming a molecule of oxygen and an atom of oxygen. However, the free atom of oxygen may recombine with another molecule of oxygen to form ozone again. On the other hand, an atom of oxygen may meet a molecule of ozone, in which case they will form two molecules of oxygen, thus destroying the ozone. In this "photochemical" region of the upper atmosphere these processes of formation and decomposition go on at the same time and there will finally be an equilibrium between the amount of ozone being formed and the amount being destroyed. Any ozone which may be carried down to the ground is immediately destroyed by contact with smoke and with vegetation. In the photochemical region there may be as much as one part of ozone by volume to every 100,000 parts of air. The ozone in the photochemical region will be slowly mixed with all the rest of the atmosphere and if there was no destruction of ozone near the ground, the whole atmosphere would contain something like this proportion of ozone, an amount that would be extremely unpleasant to breathe. While, therefore, the ozone is useful in cutting off injurious ultraviolet sunlight, it is fortunate that it is so rapidly destroyed at the ground! Because the ozone is destroyed at the ground there is a gradual drift of ozone from the photochemical region downwards through the whole of the lower atmosphere. Any ozone which comes below a height of about 20 km will be protected from the Sun's ultraviolet light by the ozone above it and will have a life of several months, provided that it does not come near the ground. Measurements made during the daytime above a grass field have shown that there is a very rapid decrease in the ozone concentration as the surface is approached, and among the grass blades there is a negligible amount of ozone. On a still night, when there is little turbulence in the lower air, the downward transport of ozone is greatly reduced and then all the air for many meters above the ground may be almost free from ozone. In smoky air it is generally impossible to detect any ozone. (The ozone which is said to be formed by sunlight on the 'smog' of Los Angeles is a peculiar case and not yet thoroughly understood.)

Ozone may be formed not only by the action of ultraviolet light but also by the action of a high-voltage electric discharge—indeed this is the way that ozone is generally produced in artificial ozonizers. We may thus expect that ozone will be formed in thunderstorms, and observations indicate that this is probably the case, though the observations are difficult and somewhat uncertain. However, the ozone produced by thunderstorms is certainly a very small part of the whole ozone in the atmosphere and is, at most, of only local importance.

3. Measurement of the Amount of Ozone in the Atmosphere

When studying the ozone in the atmosphere we generally wish to measure one of two things: (*a*) the total amount of ozone in the air above any particular place, i.e. in the whole column of air from the ground to the top of the atmosphere; (*b*) the concentration of the ozone in the air at any particular place, e.g. the concentration in the air at ground level or in the air around a balloon as it ascends through the atmosphere (from which we get the vertical distribution of ozone in the air).

A. *Total Amount of Ozone—Optical Method*

The total amount of ozone in the air above any place is most conveniently and accurately measured by observing how much certain wavelengths in sunlight are absorbed in their passage through the atmosphere. For this purpose it is usual to measure the intensity of some wavelength near the long-wave edge of the strong absorption band of ozone in the ultraviolet region. The wavelength used must be carefully chosen since, if it is too long, the absorption by ozone will be too small to give accurate values of the ozone, while if it is too short the absorption may be so strong that the light which reaches the ground may be too weak to measure accurately. In practice wavelengths around 3,050 to 3,100 Å are generally used. The solar energy in these wavelengths is reduced to about a tenth in its passage through the atmosphere, the actual amount varying with the height of the Sun and the amount of ozone present. Additional observations are made to find what would be the intensity of the light if there were no ozone present. It would be possible to use only one wavelength in the solar spectrum for these measurements but there are great advantages in measuring the *ratio* of the intensities of two adjacent wavelengths such as 3,050 and 3,250 Å, the first being strongly absorbed by ozone while the second is but little absorbed. In such a case we get nearly the full effect of the absorption at 3,050 Å

Figure 6.1
INTENSITY OF SUNLIGHT OF DIFFERENT WAVELENGTHS REACHING THE GROUND

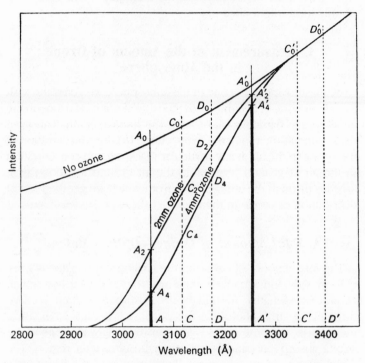

The curve A_0, A'_0, D'_0 shows the intensity of each wavelength which would reach the ground with the sun overhead if there was no ozone in the atmosphere. The curve A_2 A'_2 shows the intensity at ground level if there was 2 mm of ozone in the atmosphere, while the curve A_4 A'_4 allows for 4 mm of ozone in the atmosphere. Only the ratio of two wavelengths, e.g. $A_2 A/A'_2 A'$, is measured, not their individual intensities. Note that each additional 2 mm of ozone approximately halves the intensity of wavelength A reaching the ground, while the intensity of A' is hardly affected. Other pairs of wavelengths such as C and C', or D' and D, may be used.

while the instrumental measurements are much easier and some troublesome effects, such as the scattering of light by the air and by haze, are nearly eliminated (see Figure 6.1).

Since the absorption increases very rapidly as the wavelength decreases it is necessary to isolate the exact wavelengths required and some form of spectrograph, i.e. monochromator, must be used for accurate work. Further, the intensity of sunlight at these short wavelengths is extremely weak compared with the intensity of the visible parts of the spectrum, and great precautions must be taken in the design of the instrument to prevent the very much stronger light of longer wavelengths from producing serious errors due to scattering by all the optical surfaces in the instrument. The intensities of the required wavelengths may be measured either by photographic or photo- electric methods. In the first case the spectrum of sunlight is photographed and the ratio of the intensities of the wavelengths used is deduced from the density of their photographic images. When the photoelectric method is used the required wavelengths are isolated by two slits in the monochromator and their intensities are measured by photoelectric cells behind the slits, suitable electronic amplification being used. The amount of ozone in the atmosphere can be measured by these methods with an error of not more than 2 or 3 percent. The instrument is always used with direct sunlight when this is possible, but when sunlight cannot be used because of clouds, measurements may be made on the light from the clear zenith sky or even from the cloudy sky, though with somewhat less accuracy.

B. *Concentration of Ozone—Chemical Methods*

C. *Vertical Distribution of Ozone*

[The two sections above, which deal with various technical means for measuring ozone in the atmosphere, have been omitted.]

4. Distribution of Ozone Over the World At Different Seasons

A. *Total Ozone*

We now come to consider the results of the many thousands of measurements of atmospheric ozone that have been made all over the world. It will be convenient to give first a general, worldwide picture and then go on to the changes in the amount of ozone that occur from day to day and from place to place, and are found to be closely related to changes in the weather conditions. Figure 6.2

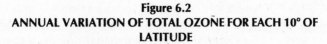

Figure 6.2
ANNUAL VARIATION OF TOTAL OZONE FOR EACH 10° OF LATITUDE

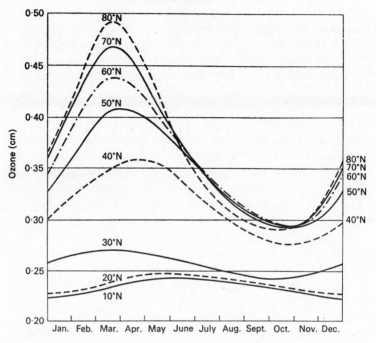

gives the annual variation of the total ozone* for each 10° of latitude in the Northern Hemisphere. This diagram is based on observations made at twenty-four places, mainly during the International Geophysical Year, and represents average conditions for all longitudes, any abnormalities being averaged out. The results of the same observations are presented again in another manner in Figure 6.3 where lines of equal ozone are plotted on a diagram with latitude as ordinates (vertically) and the month of the year as abscissae (horizontally). Unfortunately observations have not yet been made at enough places in the Southern Hemisphere to allow a similar diagram to be drawn accurately for that hemisphere,

* It is usual to express the toal amount of ozone in the atmosphere above any one place as the thickness of the same amount of ozone if it were all condensed into a layer of pure ozone at the surface of the Earth, i.e. at normal temperature and pressure.

Figure 6.3
DISTRIBUTION OF OZONE OVER THE NORTHERN HEMISPHERE
THROUGHOUT THE YEAR

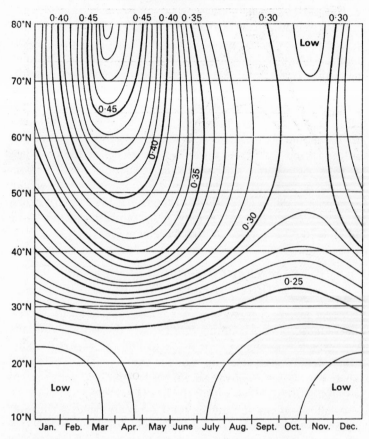

though observations near the South Pole show a most interesting abnormality in that region.

Figure 6.2 shows that all places with a latitude greater than 40°N have a large seasonal variation in the total ozone. Since the ozone is formed by sunlight, it might be expected that at these latitudes the maximum amount would be found in the late summer and the minimum in the late winter, just as the maximum and minimum of temperature are found at these seasons. However, this is far from being the case, the maximum actually being in March or April and the minimum about October. The reason for this will be discussed later. The highest ozone values of all are found in the Arctic regions during the spring, and it is important to note that even in high latitudes the amount of ozone begins to increase during December and January when the Sun is very low even at midday or may actually be below the horizon. It is also interesting to see that during the months of August to November all places with a latitude greater than about 45°N have almost the same amount of ozone, so that at this season of the year there is hardly any variation of total ozone with latitude; this is an important point to remember when we come to consider how the day-to-day variations of ozone are produced. There is a very rapid decrease in the amount of ozone between latitudes 40 and 30°N, particularly in the spring.... This is shown on Figure 6.3 by the crowding together of lines of equal ozone in these latitudes. South of about 25°N there seems to be little change of ozone with latitude or with season. It is rather surprising that the tropical belt of low ozone shows little sign of moving north and south with the Sun, being in much the same place in both December and June.

B. *Vertical Distribution of Ozone in Different Parts of the World*

Great progress has recently been made in measuring the vertical distribution of ozone in different parts of the world. While the earliest measurements were made in Europe, as has so often happened in other cases, most of our knowledge has come from an extensive series of measurements in America, with stations extending from Bolivia (16°S) to Thule (76°N).

At all places where observations have been made, at whatever latitude, little ozone has been found in the troposphere. This is to be expected since ozone is rapidly destroyed at ground level, and the large amount of mixing by turbulence in the troposphere will rapidly transfer ozone from the higher to the lower levels.

In very low latitudes the vertical distribution of ozone is fairly simple and shows little change from day to day and throughout the year. The tropopause here is, of course, always high (about 17 km) and is generally well defined by a sudden increase in temperature. On entering the stratosphere the ozone immediately begins to rise and a regular and steady increase continues up to a maximum at a height of about 25 to 27 km. Between the tropopause and the maximum, the concentration of ozone increases more

Figure 6.4
AVERAGE VERTICAL DISTRIBUTION OF OZONE IN THE ATMOSPHERE

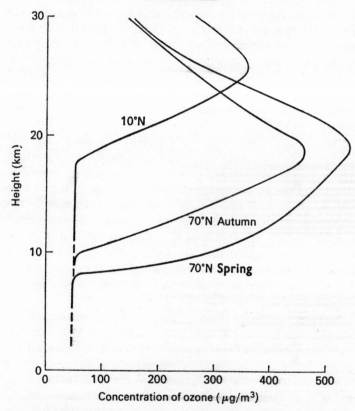

The diagram shows the average vertical distribution of ozone during the spring and autumn in high latitudes and at all seasons in low latitudes. The rate of decrease of ozone with height above about 30 km is such that the ratio of ozone to air is roughly constant.

than ten times while, owing to the decrease in the density of the air with height, the ozone/air ratio increases more than 100 times. Above about 25 to 27 km the concentration of ozone begins to decrease at much the same rate as the density of the air decreases, so that the ozone/air ratio remains nearly constant up to the greatest heights reached by balloons (about 35 km).

Measurements made in the stratosphere at higher latitudes show a more complicated structure. The level at which the maximum concentration of ozone occurs is much lower near the poles than near the equator, and shows a steady fall with latitude similar to the fall in the height of the tropopause, being only about 18 km at Thule (76°N) and at Halley Bay (75°S). Although there is always a general increase in the concentration of ozone with increasing height in the lower stratosphere, there are sometimes large irregularities, and layers of low ozone may be found above layers of higher ozone. In some cases, layers a few kilometers thick are found in which even the ozone/air ratio is less than in a lower layer.

The changes in the vertical distribution with season are very similar at all places outside the tropics. The maximum in the total ozone that is always found in spring is produced by a general increase in the concentration of ozone between the tropopause and about 20 to 25 km, with a corresponding decrease in the autumn. At high levels, above 30 km, where photochemical formation takes place, the maximum is in summer rather than in the spring, as one might expect. However, the amount of ozone at these high levels is small and the annual variation in the total ozone is mainly governed by the amount of ozone between the tropopause and 25 km.

The large variations of ozone with height that are sometimes found in the lower stratosphere, in middle and high latitudes, were unexpected, and their cause is still not well understood. They are, however, of much interest and we will therefore consider them a little more fully. As we have seen, the ozone below about 25 km is very stable, being protected from solar radiation by the ozone above it. Variations of the ozone concentration with height might be caused in one of three ways or a combination of them all.

(1) Diffusion due to turbulence will always transfer ozone from regions of high ozone/air ratio to regions of low ozone/air ratio. If there is much turbulence in any layer, the ozone/air ratio will be nearly uniform throughout the layer; conversely if there is little turbulence there may be a large difference in the ozone/air ratio between the top and the bottom of the layer. The amount of turbulence in any layer depends very largely on the fall of temperature within the layer, and we should expect to find that layers in which the temperature falls rapidly with height would be regions of high turbulence, and therefore the ozone/air ratio would in-

crease only slowly with height (e.g. the troposphere). On the other hand, layers in which the temperature changes only little with height or even increases with height, would be layers of little turbulence and might show a large difference in the ozone/air ratio between the top and the bottom.

(2) If the air is generally subsiding over a region it will bring down air having a high ozone/air ratio, while a general rising current will carry up air which is weak in ozone. If there is also some spreading sideways, this may cause a layer weak in ozone to be found above one rich in ozone.

(3) There may be cases in which tongues of air rich or weak in ozone are carried horizontally between other layers. Thus if air from the upper tropical troposphere were carried polewards into the stratosphere of higher latitudes, it would result in a layer of low ozone being sandwiched between layers of higher ozone.

Occasionally tongues of stratospheric air are drawn down into the troposphere. These cases are usually associated with polar fronts or discontinuities in the tropopause. This stratospheric air is both drier and richer in ozone than the normal tropospheric air.

C. *Cause of the Variation of Ozone with Season and Latitude*

We must now consider why the ozone in the atmosphere has this curious distribution which we have just described and why it changes in so peculiar a manner from season to season. Since the ozone is formed from oxygen by the photochemical action of sunlight at a height of some 30 km or more, the amount of ozone in this photochemical region may be expected to be greatest at those times and places where the sunlight is most intense, i.e. over the equator and also in high latitudes in summer. From this photochemical region, the ozone will be carried down by small-scale turbulence, but as the turbulence in the stratosphere is weak, owing to its very stable structure, the transport through the stratosphere will be slow. When the ozone reaches the tropopause it will be caught up in the much stronger turbulence of the troposphere and will be rapidly mixed throughout that region. The amount of ozone in the photochemical region is only about a quarter of the total ozone in the atmosphere, and the greater part of the ozone is shielded from the active wavelengths of sunlight by the ozone above it and will have a life of some months at least, though any ozone reaching the ground will be rapidly destroyed on contact with vegetation or smoke.

If there were no large-scale movements of air, the ozone distribution over the world might be expected to be similar to that in the photochemical region, but clearly this is not the case, e.g. we find the maximum amount of ozone in high latitudes in spring when, on photochemical grounds, we should expect it to be nearly at a minimum. Any large-scale up or down movements of the air will hinder or help the downward drift of ozone, while large-scale wind currents will transport ozone horizontally to different parts of the world. When looking for the cause of the very dry air in the stratosphere (Chapter 2) we had to suppose that there was a general, slow, rising current of air from the upper troposphere into the stratosphere in very low latitudes. Such a belt of rising air would also account very well for the low value of the total ozone near the equator.

Air passing upwards from the troposphere into the stratosphere in low latitudes must return to the troposphere somewhere and it is thought that this return takes place mainly in high latitudes in winter, and that a given mass of air remains in the stratosphere for a time of the order of 6 months before it returns to the troposphere. The descent of air in high latitudes in winter is probably aided by the fact that the air at heights of 15 to 40 km at this season is very cold and therefore heavy (Figure 2.2). There is probably also some return of air from the stratosphere to the troposphere in cyclonic areas of middle latitudes; it will be remembered that the upper troposphere is very dry in these regions, indicating that the air has descended from the stratosphere.

On entering the equatorial stratosphere the tropospheric air will spread out toward higher latitudes. The descending air at great heights over the winter pole will cause air to flow in from lower latitudes to take its place, and if this movement extends up to the photochemical region, ozone-rich air will be carried polewards, and as it descends it will fill the polar stratosphere with ozone-rich air. Since the tropopause here is low there is relatively much air in the stratosphere, and as this air is rich in ozone the total ozone will be high, agreeing with observation. Observations of the vertical distribution of ozone agree well with these suggestions; the ozone concentration at a height of 25 to 35 km shows an annual variation with a maximum in the summer and a minimum in the winter, while the ozone concentration in the lower layers of the stratosphere shows a maximum in the spring and a minimum in the autumn, in agreement with the total ozone.

The concentration of ozone (i.e. ozone per unit volume) is, on the average, nearly constant at all heights throughout the troposphere. Since the ozone is being transported downwards by turbulence, it might be expected that this concentration would be greater at high levels than lower down. The *ratio* of ozone to air *is* greater in the upper troposphere than below, but since the air is compressed at lower levels by the greater atmospheric pressure, there is more of both air and ozone in a given volume at low levels and this may balance the smaller ozone/air ratio. Thus the amount of ozone in a given volume may be much the same at all heights within the troposphere.

D. *Anomalies in the General Worldwide Pattern*

As already described, the ozone in nearly all parts of the world shows a marked seasonal variation with maximum in the spring and minimum in the autumn, the range (maximum minus minimum) increasing from zero at the equator, until near the North Pole the spring value is nearly twice that in the autumn. Apart from day-to-day changes, which are associated with changing meteorological conditions, the annual curve has a fairly smooth wave-form with a tendency to show a steep rise in the spring. There are two major departures from this regular worldwide pattern.

(1) Observations made at Halley Bay in the Weddell Sea (latitude 75°S), during and since the International Geophysical Year, show that the total ozone there has the usual low value in the autumn, but it does not rise much during the winter, nor even during the spring; then in November—well after the time of the expected spring maximum for the Southern Hemisphere—the ozone suddenly rises and within a week or two reaches normal values (i.e. values equal to those found at a corresponding season and latitude in the Northern Hemisphere). After this it follows the expected curve until the next autumn minimum, but, of course, it never reaches the high values found near the North Pole in March. Measurements of the upper air temperatures show that the upper part of the stratosphere at the South Pole is very cold in winter and remains cold during the spring and then suddenly, close to the time of the ozone "jump" in November, the temperature at the

highest levels rises sharply and within a few weeks may have risen 50°C. The change of temperature in the lower stratosphere is smaller and less sudden. Clearly, during November the whole structure of the south polar stratosphere has undergone a fundamental change. It seems as if in winter the south polar stratosphere is cut off from the general worldwide circulation of air by the very intense vortex of strong westerly winds which blow round the Antarctic continent, enclosing very cold air which is rather weak in ozone; neither the ozone nor the temperature rises much until this vortex suddenly breaks down in November. However, much further work will be necessary before these conditions are fully understood.

(2) The second anomaly is found in high latitudes in Canada. Here sudden rises both in the temperature of the stratosphere and in the total ozone are also found, but while near the South Pole the change has so far always taken place in November, in north Canada the rise may be found at any time between January and April. At Spitzbergen the seasonal variation seems to be more normal, though the rise in ozone during January and February is generally rapid. Clearly neither the sudden warming nor the increase in ozone can be due directly to solar radiation, for in Canada the rise can occur at times when the Sun is still below the horizon and in no case could the solar heating produce so rapid a warming. Both the warming and the increase in ozone are almost certainly due to descending air in the upper stratosphere which causes a warming of the air by compression and carries down ozone-rich air into the lower stratosphere. Another abnormality appears to exist in North India where ozone values seem to be consistently lower than those at similar latitudes in America, Africa, and Japan.

E. *Fall-out from Thermonuclear Explosions*

A rather striking confirmation of the circulation of the air in the stratosphere, which was suggested by the study of atmospheric ozone and humidity, has recently come from measurements of the fall-out of radioactive materials from thermonuclear explosions. While the earlier atomic explosions projected radioactive matter into the upper troposphere, from where it fell out, or was washed out, in a matter of weeks, the larger, thermonuclear explosions projected radioactive matter right up into the upper stratosphere from where it only slowly returned to the Earth. The fall-out of the long-lived radioactive elements has been measured in many different parts of the world and its variations show striking resemblances to those of atmospheric ozone. In the first place, although many of the explosions took place in low northern latitudes, most of the fall-out was found in middle latitudes and very little radioactive matter came down near the equator. It is remarkable that even the small amount of radioactive material which drifted across the equator into the Southern Hemisphere fell out in greater quantities in middle southern latitudes than nearer the equator and in spring rather than in autumn. The general rising current of air in the upper troposphere and lower stratosphere, which we have suggested exists in low latitudes, well accounts for the low fall-out in this part of the world. In view of the fact that the ozone is formed continuously at high levels all over the sunlit part of the world, while radioactive material is injected in isolated places and at definite times, it is not surprising that there is a difference in the latitude of the maximum ozone and the maximum fall-out, the ozone having a maximum in high latitudes and the fall-out in middle latitudes.

The second point of interest is that the rate of fall-out in temperate latitudes shows a strong maximum in the spring (possibly slightly later than the ozone maximum) and a minimum in the autumn, almost exactly in line with the seasonal variation of ozone. Unfortunately there are few measurements of the radioactive fall-out which can be used to see whether it also follows the changes in atmospheric ozone with varying meteorological conditions, which we shall discuss in the next section, but some measurements indicate that this is the case.

5. Ozone and Weather—Day-to-Day Variations

A. (i) *Middle Latitudes*

Up to now we have considered the ozone values averaged over a period of a month or more and we now turn to the much more rapid variations which take place within a few days. In the same way we have previously considered only average values over wide areas of the world, averaging out the variations which are found on many days between places a few hundred miles apart. Actually the rapid, and relatively local, variations of ozone are as large as the seasonal and worldwide variations, so that on a day with relatively low ozone in spring, the ozone value may be the same as on a day with relatively high ozone in the autumn. These rapid, local variations are of great interest since they are found to be closely associated with other meteorological conditions. Since most of the ozone is known to exist above the tropopause, it is only to be expected that these short-period variations would be more closely related to the meteorological conditions in the upper atmosphere than to the surface conditions and this is found to be the case. It has been shown in Chapter 2 that there are close connections between the variations of the temperature, the pressure, the height of tropopause, and other variables in the upper atmosphere, so that we naturally find that the variations in ozone are connected with all these to a greater or less extent. There is a tendency for the following associations.

High Ozone	*Low Ozone*
Cyclonic wind circulation at the level of the tropopause	Anticyclonic wind circulation at the tropopause
High temperature in the stratosphere	Low temperature in the stratosphere
Low temperature in the troposphere	High temperature in the troposphere
Low level of tropopause	High level of tropopause
Low absolute pressure	High absolute pressure

These relations are brought out in Figure 6.5 where 3-day running means of the ozone, the height of the 200-mb surface, and the height of the tropopause are plotted one above the other. It is found that if 3-day running mean values are used, the curves show a rather closer connection than if individual daily values are plotted, probably because small errors of observation are reduced but the real changes are not smoothed out too much.

While the variations in the amount of ozone are associated with changes in the height of the tropopause, etc., the closest relation of all is shown by the *type* of pressure distribution at about the height of the tropopause. On those days when the pressure maps

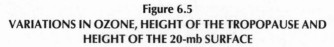

Figure 6.5
VARIATIONS IN OZONE, HEIGHT OF THE TROPOPAUSE AND
HEIGHT OF THE 20-mb SURFACE

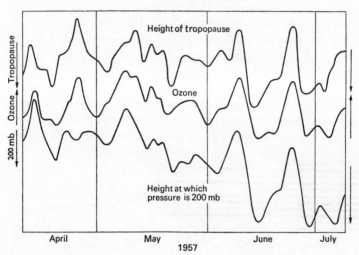

The curves show 3-day running means of the ozone at Oxford, and also the height of the tropopause and the height of the 200-mb surface at Crawley (100 km SE. of Oxford). The top and bottom curves are plotted with heights increasing downwards to conform with the variations of ozone. The height of the tropopause and the 200-mb surface are taken from the Aerological Record, Meteorological Office, Air Ministry, by kind permission of the Director-General.

for a height of about 9 km (300 mb) show cyclonic conditions, the amount of ozone is relatively large, while when the conditions are anticyclonic there is relatively little ozone. The 300-mb weather maps for three typical pairs of days are shown in Figure 6.6 *(a)*, *(b)*, and *(c)*. In each figure two contrasted days are shown, one having cyclonic and the other anticyclonic conditions. The maps give the contours of the 300-mb surface, the contours being very similar to isobars.

The absolute change in the amount of ozone between cyclonic and anticyclonic days in spring is rather greater than the corresponding change in the autumn, but since the average value of the ozone is lower in the autumn the percentage changes in spring and autumn are not very different. For those readers who are more used to seeing the weather maps for the ground level than for the upper air, it may be well to point out that systems of closed isobars

Figure 6.6
PRESSURE DISTRIBUTION AT THE 300-mb LEVEL FOR CYCLONIC AND ANTICYCLONIC DAYS

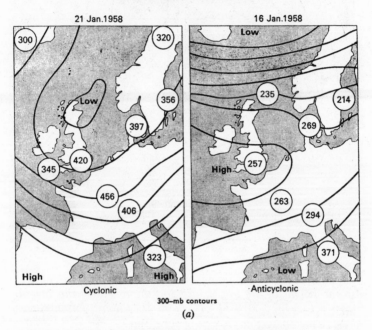

Cyclonic   Anticyclonic

300–mb contours

(a)

These three pairs of maps show the pressure distribution at a height of about 9 km by means of contours of the 300-mb surface. The figures within the circles give the values of the total ozone at each place in thousandths of a centimeter. Each pair of days contrasts cyclonic and anticyclonic conditions. The high value of the ozone in cyclonic conditions is clearly seen. The contours of the 300-mb surface are reproduced from the Aerological Record, Meteorological Office, Air Ministry, by kind permission of the Director-General, as are also the ozone values for Lerwick, Eskdalemuir, and Cambourne.

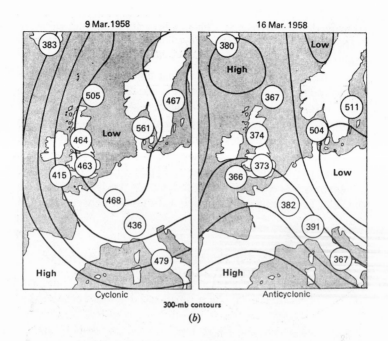

300-mb contours

(b)

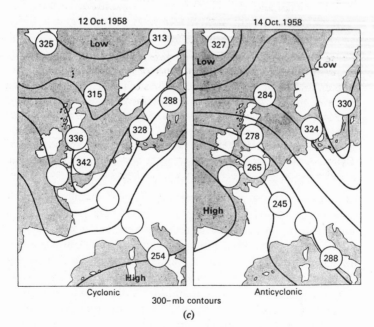

300-mb contours

(c)

showing definite areas of low pressure (cyclones) or high pressure (anticyclones) at low levels generally tend to become troughs of low pressure and ridges of high pressure in the upper atmosphere. It will be seen from these maps that the relation between the ozone and the pressure distribution is equally close at all times of the year (an important point to which we shall refer again later). It will also be seen that in the transition region between a trough and a ridge, the changes in ozone may be very rapid. On 16 March 1958, for example, the value was 0.504 cm over Denmark (cyclonic) and only 0.373 cm over England (anticyclonic). As a trough moves across the country followed by a ridge, or vice versa, one may find changes of as much as 0.100 cm between one day and the next.

(ii) *Day-to-day Changes in Tropical Regions*

It has already been pointed out that in low latitudes the total amount of ozone is everywhere small and that there is little variation from day to day. The storms that occur in low latitudes are mostly confined to the lower levels of the atmosphere and, as would be expected, do not cause any change in the total ozone. Even the change from SW to NE monsoon and vice versa in India is not accompanied any appreciable change in the amount of ozone, since the monsoon does not affect the upper atmosphere. In north India during the winter months depressions may move in from middle latitudes and these give an increase in the amount of ozone just as depressions do in higher latitudes.

(iii) *Day-to-day Changes in Polar Regions*

Until the beginning of the International Geophysical Year (July 1957) the only measurements of the amount of ozone in very high latitudes were those at Spitzbergen (78°N) and, owing to the very difficult conditions there, the number of observations was rather small. During the I.G.Y. measurements were made at Halley Bay (75°S), at Alert (82.5°N), and at Resolute (75°N). Daily measurements at these stations can, of course, only be made during the summer half of the year but these indicate that the day-to-day changes are similar to those found in middle latitudes and show much the same relation to the other meteorological conditions as is found in middle latitudes. When all the measurements made during the I.G.Y. have been studied we may hope to know more about the changes in high latitudes.

B. *The Cause of the Connection Between Ozone And Other Upper Air Conditions*

We must now consider what causes the amount of ozone in the upper atmosphere to change from day to day, and why the variations are closely connected with other meteorological conditions. As with the changes in total ozone between spring and autumn, the changes between cyclonic and anticyclonic conditions are found to take place mainly in the first 5 to 10 km above the tropopause. A cyclonic depression, shown on the surface weather map as a closed low pressure area, is represented at a height of 15 km by a trough of low pressure extending to lower latitudes; on the other hand, an anticyclone, shown on the surface weather map as an area of high pressure is represented at 15 km by a ridge of high pressure extending towards the pole. These troughs and ridges tend to circulate round the pole from west to east, but the general westerly wind at these heights has a much greater speed, and the air actually flows *through* these troughs and ridges. As the air blows into a low pressure trough it descends, while as it approaches a ridge it ascends. These descending and ascending movements of the air will lead to increases and decreases in the amount of ozone as we have described a little earlier, and will, to some extent at least, account for the greater amount of ozone in depressions than in anticyclones. In spring—but not in autumn— the general amount of ozone is greater in high latitudes than in low latitudes, so that north polar currents will carry southward air which is rich in ozone, while equatorial currents will tend to carry polewards air which is weak in ozone. This is also a cause in producing changes in the total ozone. However, in autumn, as shown in Figures 6.2 and 6.3, there is little change in the amount of total ozone found at any latitude north of about 45°N so that at this season the effect will not apply. We have still much to learn about these changes in ozone.

Index

Rogelio A. (Roger) Maduro, *is an associate editor of 21st Century Science & Technology magazine and the author of many articles on Earth and atmospheric sciences, environment, environmental legislation, and new developments in industrial technologies. A native of Panama, Maduro studied civil engineering at Washington University in St. Louis and earned a bachelor of science degree in geology from The City College of New York.*

As a researcher for the Schiller Institute, Maduro wrote two chapters of a 1986 book The Integration of Ibero-America, *on infrastructure development projects (transportation, hydroelectric, and nuclear) and on large-scale mining and industrial complexes for the continent.*

He is married with three children and resides in Virginia.

Ralf Schauerhammer, *a native of Jena, Germany, is chairman of the Fusion Energy Forum and editor of its German-language journal,* Fusion—Technology and Science for the 21st Century. *He has authored numerous articles on science and environmentalism, and his 1991 book* Sackgasse Oekostaat, kein Platz fuer Menschen *(Dead end—The environmental state: No room for human beings), is a best-seller in Germany.*

Schauerhammer studied mathematics and physics at the University of Mainz. Since 1979, he has worked with an international task force on the simulation of economic development according to the LaRouche-Riemann model. In 1980, he coauthored the book Industrialization of Africa *(1980).*